v

2

Yakov G. Sinai

Probability Theory

An Introductory Course

Translated from the Russian
by D. Haughton

With 14 Figures

Springer-Verlag

Berlin Heidelberg New York
London Paris Tokyo
Hong Kong Barcelona
Budapest

Yakov G. Sinai
Russian Academy of Sciences
L. D. Landau Institute for Theoretical Physics
ul. Kosygina, 2
Moscow 117940, GSP-1, V-334, Russia

Translator:

Dominique Haughton
Department of Mathematics
College of Pure and Applied Science
University of Lowell, One University Avenue
Lowell, MA 01854, USA

Title of the Russian original edition (in two parts):
Kurs teorii veroyatnostej
Publisher MGU, 1985 and 1986

Mathematics Subject Classification (1991): 60-01

ISBN 3-540-53348-6 Springer-Verlag Berlin Heidelberg New York
ISBN 0-387-53348-6 Springer-Verlag New York Berlin Heidelberg

Die Deutsche Bibliothek – CIP-Einheitsaufnahme.
Sinai, Yakov G.: Probability theory: an introductory course / Yakov G. Sinai. Transl. by
D. Haughton. – Berlin; Heidelberg; New York; London; Paris; Tokyo; Hong Kong;
Barcelona; Budapest: Springer, 1992 (Springer textbook) Einheitssacht.: Kurs teorii vero-
jatnostej < engl. > ISBN 3-540-53348-6 (Berlin . . .) ISBN 0-387-53348-6 (New York . . .)

Typesetting: Camera ready by author
41/3140 - 5 4 3 2 1 0   Printed on acid-free paper

Prefaces

Preface to the English Edition

This book grew out of lecture courses that were given to second and third year students at the Mathematics Department of Moscow State University for many years. It requires some knowledge of measure theory, and the modern language of the theory of measurable partitions is frequently used. In several places I have tried to emphasize some of the connections with current trends in probability theory and statistical mechanics.

I thank Professor Haughton for her excellent translation.

Ya. G. Sinai   May 1992

Preface to First Ten Chapters

The first ten chapters of this book comprise an extended version of the first part of a required course in Probability Theory, which I have been teaching for may years to fourth semester students in the Department of Mechanics and Mathematics at Moscow State University. The fundamental idea is to give a logically coherent introduction to the subject, while making as little use as possible of the apparatus of measure theory and Lebesgue integration. To this end, it was necessary to modify a number of details in the presentation of some long-established sections.

These chapters cover the concepts of random variable, mathematical expectation and variance, as well as sequences of independent trials, Markov chains, and random walks on a lattice. Kolmogorov's axioms are used throughout the text. Several non-traditional topics are also touched upon, including the problem of percolation, and the introduction of conditional probability through the concept of measurable partitions. With the inclusion of these topics it is hoped that students will become actively involved in scientific research at an early stage.

This part of the book was refereed by B.V. Gnedenko, Member of the Academy of Science of the Ukrainian Socialist Soviet Republic, and N.N. Chentsov, Doctor of Mathematical and Physical Sciences. I wish to thank I.S. Sineva for assistance in preparing the original manuscript for publication.

Preface to Chapters 11 - 16

Chapters eleven through sixteen constitute the second part of a course on Probability Theory for mathematics students in the Department of Mechanics and Mathematics of Moscow State University. The chapters cover the strong law of large numbers, the weak convergence of probability distributions, and the central limit theorem for sums of independent random variables. The notion of stability, as it relates to the central limit theorem, is discussed from the point of view of the method of renormalization group theory in statistical physics, as is a somewhat less traditional topic, as is the analysis of asymptotic probabilities of large deviations.

This part of the book was also refereed by B.V. Gnedenko, as well as by Professor A.N. Shiryaev. I wish to thank M.L. Blank, A. Dzhalilov, E.O. Lokutsievckaya and I.S. Sineva for their great assistance in preparing the original manuscript for publication.

Translator's Preface

The Russian version of this book was published in two parts. Part I, covering the first ten chapters, appeared in 1985. Part II appeared in 1986, and only the first six chapters have been translated here.

I would like to thank Jonathan Haughton for typesetting the English version in TeX.

Contents

Lecture 1. Probability Spaces and Random Variables

1.1 Probability Spaces and Random Variables

The field of probability theory is somewhat different from the other fields of mathematics. To understand it, and to explain the meaning of many definitions, concepts and results, we need to draw on real life examples, or examples from related areas of mathematics or theoretical physics. On the other hand, probability theory is a branch of mathematics, with its own axioms and methods, and with close connections with some other areas of mathematics, most notably with functional analysis, and ordinary and partial differential equations. At the foundation of probability theory lies abstract measure theory, but the role played here by measure theory is similar to that played by differential and integral calculus in the theory of differential equations.

We begin our study of probability theory with its axioms, which were introduced by the great Soviet mathematician and founder of modern probability theory, A. N. Kolmogorov. The first concept is that of the space of elementary outcomes. From a general point of view the space of elementary outcomes is an abstract set. In probability theory, as a rule, it is denoted by Ω and its points are denoted by ω, and are called elementary events or elementary outcomes. When Ω is finite or countable, probability theory is sometimes called discrete probability theory. A major portion of the first half of this course will be related to discrete probability theory. Subsets $C \subset \Omega$ are called events. Moreover the set $C = \Omega$ is call the certain event, and the set $C = \emptyset$ (sometimes written as $C = \mathcal{N}$) is called the impossible event. One can form unions, intersections, Cartesian products, complements, differences, etc. of events.

Our first task is that of constructing spaces of elementary outcomes for various concrete situations, and this is best done by considering several examples.

Example 1.1. At school many of us have heard that probability theory originated in the analysis of problems arising from games of chance. The simplest, although important, example of such a game consists of a single toss of a coin, which can fall on one of its two sides. Here Ω consists of two points, $\Omega = (\omega_h, \omega_t)$, where ω_h (ω_t) means that the toss gave heads (tails).

Remark. In this example the number of possible events $C \subset \Omega$ is equal to $4 = 2^2$: $\Omega, \omega_h, \omega_t, \emptyset$.

If we replace the coin by a die then Ω consists of six points, $\omega_1, \omega_2, \omega_3,$ $\omega_4, \omega_5, \omega_6$, according to the number showing on the top side after throwing the die.

Example 1.2. Let us assume that a coin is tossed n times. An elementary outcome then has the form $\omega = \{a_1, a_2, \ldots, a_n\}$. Each a_i takes two values, 1 or 0. We will say that $a_1 = 1$ if the toss yielded heads, and $a_i = 0$ if the toss yielded tails. Alternatively we could say that ω is a word of length n, written from the alphabet $\{0, 1\}$. The space Ω in this case consists of 2^n points.

This space Ω is encountered in many problems of probability theory. For example one can imagine a complicated system consisting of n identical elements, each of which can be found in two states, which as before will be denoted by $\{0, 1\}$. The state of the whole system can then be described by a word ω, as above.

In statistical mechanics models, one encounters models which consist of individual elementary magnets (spins) with different orientations. We represent them by an arrow directed upwards (downwards) if the north pole is above (below):

Fig. 1.1.

Then a collection of magnets situated at the points $1, \ldots, n$ is a collection of arrows

↑	↑	↓	↑	↓	↓	↓		↑	↓	↓	↑
1	1	0	1	0	0	0	...	1	0	0	1

which we can again codify by a word of length n from $(0, 1)$.

Several generalizations suggest themselves immediately. First we can assume that each a_i takes not two, but several values. In general it is convenient to assume that a_i is an element of an abstract space of values, X. We can represent X as an alphabet, and $\omega = \{a_1, \ldots, a_n\}$ as a word of length n, written from the alphabet X. If $X = (x^{(1)}, \ldots, x^{(r)})$ contains r elements, then each letter a_i takes r values and Ω consists of r^n points. Imagine, for example, the following concrete situation. Consider a group of n students, whose birthdays are known. Assume that none of the participants were born in a leap year. Then $\omega = \{a_1, \ldots, a_n\}$ is the list of n birthdays, where each a_i takes one of

365 possible values. By the same token X is the space of all possible birthdays, and therefore consists of 365 points, and Ω consists of 365^n points. If X is the unit interval, $X = [0, 1]$, then Ω consists of all choices of n numbers on the interval $[0, 1]$. Sometimes the space Ω is written as $\Omega = X^{[1,n]}$, where $[1, n] = \{1, 2, 3, \ldots, n\}$.

The second generalization concerns a more general interpretation of the index i in the notation $\omega = \{a_i\}$. For example we could have $i = (i_1, i_2)$, where i_1 and i_2 are integers, $K_1 \leq i_1 \leq K_2$, $L_1 \leq i_2 \leq L_2$. Denote by I the set of points $i = (i_1, i_2)$ of the plane with integer coordinates.

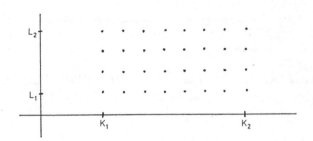

Fig. 1.2.

Further assume that each point i can be found in two states, denoted by 0,1. The state of the whole system is then $\omega = \{a_i\}$, $i \in I$, and a_i takes the two values 0,1. Such spaces Ω are encountered in the theory of random fields. X can also be arbitrary.

We now come to a general definition.

Definition 1.1. Let X be a space of values and I be a space of indices. The space $\Omega = X^I$ is the space of all possible words $\omega = \{a_i, i \in I\}$, where $a_i \in X$. Sometimes Ω is called the space of functions from I with values in X.

Example 1.3. In the lottery "6 from 36" we consider a square with 36 cells numbered 1 through 36. Each participant must choose six distinct numbers. Here an individual elementary outcome has the form $\omega = \{a_1, a_2, a_3, a_4, a_5, a_6\}$ and is a subset of six elements of the set with 36 elements. Now Ω consists of $\binom{36}{6} = (36.35.34.33.32.31)/6!$ points.

We can obtain a variant in the following way. Let X be a finite set and let us carry out a sampling without replacement of n elements of this set, i.e. $\omega = \{a_1, a_2, \ldots, a_n\}$, where now $a_i \neq a_j$. If $|X| = r$ then Ω consists of $r(r-1)\ldots(r-n+1)$ elements.[1]

---

[1] From now on the absolute value of a set will denote its cardinality, that is the number of its elements. Other examples of spaces Ω will be given later.

Let Ω be finite, i.e. $|\Omega| = m < \infty$. We denote by $\mathcal{F}$ the collection of all events C, i.e. all subspaces of Ω .

Theorem 1.1. *$\mathcal{F}$ consists of 2^m events.*

Proof. Let $C \subset \Omega$. We introduce the function $\chi_C(\omega)$ on Ω, where

$$\chi_C(\omega) = \begin{cases} 1, & \text{if } \omega \in C; \\ 0, & \text{otherwise.} \end{cases}$$

It is clear that every such function on Ω taking the values 1 and 0 uniquely defines an event C, namely the set C where the function $\chi_C(\omega)$ is equal to 1. In other words, $\mathcal{F}$ is in one to one correspondence with the points of X^{Ω}, where $X = \{1, 0\}$. The number of such points is equal to $2^{|\Omega|} = 2^m$.   □

Now let Ω be arbitrary.

Definition 1.2. A collection $\mathcal{G}$ of subsets of Ω is called an algebra if

1) $\Omega \in \mathcal{G}$;
2) $C \in \mathcal{G} \Rightarrow \Omega \setminus C \in \mathcal{G}$;
3) $C_1, C_2, \ldots, C_k \in \mathcal{G} \Rightarrow \bigcup_{i=1}^{k} C_i \in \mathcal{G}$.

Here are some simple corollaries of this definition.

Corollary 1.1. $\emptyset \in \mathcal{G}$.

Proof. Take $\Omega \in \mathcal{G}$ and apply 2).   □

Corollary 1.2. $C_1, C_2, \ldots, C_k \in \mathcal{G} \Rightarrow \bigcap_{i=1}^{k} C_i \in \mathcal{G}$.

Proof. Indeed $\Omega \setminus \bigcap_{i=1}^{k} C_i = \bigcup_{i=1}^{k} (\Omega \setminus C_i) \in \mathcal{G}$. Consequently $\bigcap_{i=1}^{k} C_i \in \mathcal{G}$.   □

The following result is already less trivial.

Theorem 1.2. *If the algebra $\mathcal{G}$ is finite, then there exist subsets $A_1, A_2, \ldots$ $\ldots, A_r$ such that 1) $A_i \cap A_j = \emptyset$ for $i \neq j$; 2) $\bigcup_{i=1}^{r} A_i = \Omega$; 3) any event $C \in \mathcal{G}$ is a union of A_i's.*

Remark. The collection of events $A_1, \ldots, A_r$ defines a partition of Ω. This way finite algebras are generated by finite partitions, and conversely. This statement can be generalized considerably.

Remark. The algebra $\mathcal{G}$ contains 2^r elements. Indeed, let us define a new space Ω' whose points are the sets in the partition $A_1, A_2, \ldots, A_r$. Then $\mathcal{G}$ can be considered as the collection of subsets of the space Ω', $|\Omega'| = r$. By Theorem 1.1 $\mathcal{G}$ contains 2^r elements.

Proof of Theorem 1.2. Let us number the elements of $\mathcal{G}$ in an arbitrary way:

$$\mathcal{G} = \{C_1, \dots, C_s\}.$$

For any C we set

$$C^1 = C, \qquad C^{-1} = \Omega \setminus C.$$

Consider the sequence $b = \{b_1, b_2, \dots, b_s\}$, where each b_i takes the value $+1$ or -1, and set

$$C^{(b)} = \bigcap_{i=1}^{s} C_i^{b_i}.$$

Then all the $C^{(b)} \in \mathcal{G}$ (as follows from 2) and Corollary 1.2).

It is possible that $C^{(b)} = \emptyset$. However for any $\omega \in \Omega$ there exists a b such that $\omega \in C^{(b)}$. Indeed $\omega \in C_i^{b_i}$ for one of the values $b_i = \pm 1$ for every i, $1 \le i \le s$. Therefore $\omega \in C^{(b)}$, and so not all $C^{(b)}$ are empty. If $b' \ne b''$ then $C^{(b')} \cap C^{(b'')} = \emptyset$. Indeed $b' \ne b''$ means that $b_i' \ne b_i''$ for some i. To fix ideas, assume that $b_i' = 1$, $b_i'' = -1$. Then in the expression $C^{(b')}$ we find $C_i^1 = C_i$, so $C^{(b')} \subseteq C_i$, and in the expression $C^{(b'')}$ we find $C_i^{-1} = \Omega \setminus C_i$, therefore $C^{(b'')} \subseteq \Omega \setminus C_i$; it follows that $C^{(b')} \cap C^{(b'')} = \emptyset$. Therefore the sets $C^{(b)}$ are empty or non-empty and they are disjoint. Let us take, for A_i, the non-empty intersections. Since each ω is an element of one of the $C^{(b)}$, it belongs to one of the A_i. This means that $\bigcup_i A_i = \Omega$.   □

Definition 1.3. A collection $\mathcal{F}$ of subsets of Ω is called a $\sigma-$algebra if $\mathcal{F}$ is an algebra and if $C_1, \dots, C_k, \dots \in \mathcal{F}$ implies $\bigcup_{i=1}^{\infty} C_i \in \mathcal{F}$.

As above, we have:

Corollary 1.3. *if $C_1, C_2, \dots, C_k, \dots \in \mathcal{F}$, then $\bigcap_i C_i \in \mathcal{F}$.*

At first sight it seems that the difference between algebras and $\sigma-$algebras is not very large. However this is not the case. In fact, only in the case of $\sigma-$algebras does an interesting theory arise.

Definition 1.4. A measurable space is a pair $(\Omega, \mathcal{F})$ where $\mathcal{F}$ is a $\sigma-$algebra of subsets of the space Ω.

We now consider the case of a discrete space Ω.

Definition 1.5. Any function $f = f(\omega)$ with real values is called a random variable.

For more general spaces Ω, the function f must have further properties, known as measurability properties (see below).

Notation. We denote random variables by small greek letters, for example ξ, η and ς. If an emphasis is needed on the functional dependence on ω then we write $\xi(\omega)$, $\xi = f(\omega)$, and so on.

The random variables form a linear space, since they can be added and multiplied by constant factors. This space is also a commutative ring, since random variables can be multiplied.

We denote by $\mathcal{F}$ the σ–algebra of all the subsets of Ω (Ω is discrete!), and we consider an arbitrary sub-algebra $\mathcal{F}_0 \subset \mathcal{F}$.

Definition 1.6. A random variable $\xi = f(\omega)$ is said to be $\mathcal{F}_0$-measurable if for any a, b, $a \le b$, the set $\{\omega : a \le f(\omega) < b\} \in \mathcal{F}_0$.

In order to understand the meaning of this concept, consider the case where $\mathcal{F}_0$ is finite. By Theorem 1.2 such an $\mathcal{F}_0$ gives rise to a finite partition of Ω into subsets $A_1, A_2, \ldots, A_r$.

Theorem 1.3. *If ξ is $\mathcal{F}_0$–measurable, it takes constant values on each A_i, $1 \le i \le r$.*

Proof. Let d_i be the different values of the function f. Their number does not exceed the number of elements of $\mathcal{F}_0$ (check this!), and therefore is finite. Let us write them in increasing order: $d_1 < d_2 < \ldots < d_\ell$, and let $U_i = \{\omega : f(\omega) = d_i\}$. It is clear that $U_i \cap U_j = \emptyset$ for $i \ne j$ and $\bigcup_i U_i = \Omega$. We now show that $U_i \in \mathcal{F}_0$. Let us construct intervals $[a_i, b_i)$ where $d_{i-1} < a_i < d_i < b_i < d_{i+1}$. Then

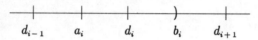

$$d_{i-1} \qquad a_i \qquad d_i \qquad b_i \qquad d_{i+1}$$

$\{\omega : a_i \le f(\omega) < b_i\} = U_i$, that is $U_i \in \mathcal{F}_0$. By Theorem 1.2, U_i can be written as a union of A_j's, i.e. $f(\omega) = d_i$ for $\omega \in A_j$. Moreover different i give rise to different elements of the partition. $\square$

We now return to the general case and turn our attention to the central concept of our theory, the concept of probability. We noted earlier that probability is a special case of measure. Now consider an arbitrary measurable space $(\Omega, \mathcal{F})$.

Definition 1.7. A probability measure is a function P, defined on $\mathcal{F}$, which satisfies the following conditions:

1) $P(C) \ge 0$ for any $C \in \mathcal{F}$;
2) $P(\Omega) = 1$;
3) if $C_i \in \mathcal{F}$, $i = 1, 2, \ldots$ and $C_i \cap C_j = \emptyset$, then

$$P\left(\bigcup_{i=1}^{\infty} C_i\right) = \sum_{i=1}^{\infty} P(C_i).$$

The number $P(C)$ is called the probability of the event C.

We now discuss the meaning of properties 1) – 3). The concrete idea of probability of an event or a phenomenon or an occurrence is the frequency of occurrence of this event, or the "chances" of its taking place. Since the frequency of occurrence is always non-negative, a probability must be non-negative. Property 2) means that when we consider an experiment whose outcome can be any point $\omega \in \Omega$ we are dealing with a situation where some outcome of the experiment will be observed. Sometimes property 2) is called the normalization property. Those events C for which $P(C) = 1$ are called certain (they have a 100% chance of occurrence). Since $\Omega = \Omega \cup \emptyset$ and $\Omega \cap \emptyset = \emptyset$, $P(\Omega) = P(\Omega \cup \emptyset) = P(\Omega) + P(\emptyset)$, and so $P(\emptyset) = 0$. Those events C for which $P(C) = 0$ are called impossible (0% chances of occurrence). The meaning of property 3) will be developed gradually. It is called the property of countable additivity, or $\sigma-$additivity of probability measures. It is fundamental to general measure theory. Properties 1)–3) show that probabilities are normed measures.

Definition 1.8. The triplet $(\Omega, \mathcal{F}, P)$ is called a probability space.

We now introduce simple corollaries to Definition 1.8. Let Ω be discrete, that is $\Omega = \{\omega_i\}$. We denote by C_ω the event with only one point: $C_\omega = \omega \subset \Omega$. Then $p(\omega) = P(C_\omega) = P(\omega)$ is the probability of the elementary outcome ω. It follows easily from properties 1)–3) of a probability that:

$$1) \quad p(\omega) \geq 0, \qquad 2) \quad \sum_\omega p(\omega) = 1. \tag{1.1}$$

A collection of numbers which satisfy 1) and 2) is called a probability distribution. In this way every probability P generates a probability distribution on Ω, and vice versa.

Theorem 1.4. *Every probability distribution on Ω generates a probability measure on the $\sigma-$algebra $\mathcal{F}$ of subsets of Ω by the formula $P(C) = \sum_{\omega_k \in C} p(\omega_k)$.*

Proof. Properties 1) and 2) of Definition 1.6 follow from (1.1). Property 3) involves

$$\sum_i P(C_i) = \sum_i \sum_{\omega_k \in C_i} p(\omega_k)$$

and follows from the fact that the sum of a converging series with positive terms does not depend on the order of summation. $\square$

Here are several equivalent formulations of $\sigma-$additivity.

Theorem 1.5. *A function P on $\mathcal{F}$ which satisfies properties 1) and 2) of Definition 1.6, and is additive (i.e. property 3 of Definition 1.6 holds for finite collections C_i) is $\sigma-$additive (satisfies property 3) if and only if*

1) for any sequence of events $C_i \in \mathcal{F}$, $C_i \subseteq C_{i+1}$,

$$P\left(\bigcup_i C_i\right) = \lim_{i \to \infty} P(C_i);$$

2) for any sequence of events $C_i \in \mathcal{F}$, $C_i \supseteq C_{i+1}$,

$$P\left(\bigcap_i C_i\right) = \lim_{i \to \infty} P(C_i);$$

3) for any sequence of events $C_i \in \mathcal{F}$, $C_i \supseteq C_{i+1}$, $\bigcap_i C_i = \emptyset$,

$$\lim_{i \to \infty} P(C_i) = 0.$$

Proof. All three statements are proved in the same way. By way of example let us prove the last one. Let $C_i \in \mathcal{F}$, $C_i \supseteq C_{i+1}$, $\bigcap_i C_i = \emptyset$. Now form the events $B_i = C_i \setminus C_{i+1}$. Then $B_i \cap B_j = \emptyset$ for $i \neq j$, $C_k = \bigcup_{i \geq k} B_i$. From the $\sigma-$additivity property of Definition 1.6 we have $P(C_1) = \sum_{i=1}^{\infty} P(B_i)$. Therefore the remainder of the series $\sum_{i=k}^{\infty} P(B_i) = P(C_k) \to 0$ as $k \to \infty$. Conversely, assume that we have a sequence of sets C_i, $C_i \cap C_j = \emptyset$, for $i \neq j$. Form $C = \bigcup_{i=1}^{\infty} C_i$. Then $C = \bigcup_{i=1}^{n} C_i \cup \bigcup_{i=n+1}^{\infty} C_i$ for any n, and by finite additivity $P(C) = \sum_{i=1}^{n} P(C_i) + P(\bigcup_{i=n+1}^{\infty} C_i)$. The events $B_n = \bigcup_{n+1}^{\infty} C_i$ decrease, and $\bigcap_n B_n = \emptyset$. Therefore $P(B_n) \to 0$ and $P(C) = \sum_{i=1}^{\infty} P(C_i)$. $\square$

Corollary to the additivity of P. *For any collection of sets $A_1, \ldots, A_n \in \mathcal{F}$, $P(\bigcup_{i=1}^{n} A_i) \leq \sum_{i=1}^{n} P(A_i)$.*

Proof. For $n = 1$ the statement is clear. We proceed by induction. We have

$$\bigcup_{i=1}^{n} A_i = \bigcup_{i=1}^{n-1} A_i \cup A_n = \bigcup_{i=1}^{n-1} A_i \cup \left(A_n \setminus \bigcup_{i=1}^{n-1} A_i\right).$$

The sets $\bigcup_{i=1}^{n-1} A_i$ and $A_n \setminus \bigcup_{i=1}^{n-1} A_i$ do not intersect. Therefore

$$P\left(\bigcup_{i=1}^{n} A_i\right) = P\left(\bigcup_{i=1}^{n-1} A_i \cup \left(A_n \setminus \bigcup_{i=1}^{n-1} A_i\right)\right)$$

$$= P\left(\bigcup_{i=1}^{n-1} A_i\right) + P\left(A_n \setminus \bigcup_{i=1}^{n-1} A_i\right)$$

$$\leq \sum_{i=1}^{n-1} P(A_i) + P\left(A_n \setminus \bigcup_{i=1}^{n-1} A_i\right)$$

using the induction hypothesis. Furthermore $A_n = (A_n \setminus \bigcup_{i=1}^{n-1} A_i) \cup (A_n \cap (\bigcup_{i=1}^{n-1} A_i))$ and by additivity of the probability

$$P(A_n) = P\left(A_n \setminus \left(\bigcup_{i=1}^{n-1} A_i\right)\right) + P\left(A_n \cap \left(\bigcup_{i=1}^{n-1} A_i\right)\right) \geq P\left(A_n \setminus \bigcup_{i=1}^{n-1} A_i\right).$$

Finally we obtain

$$P\left(\bigcup_{i=1}^{n} A_i\right) \leq \sum_{i=1}^{n-1} P(A_i) + P\left(A_n \setminus \bigcup_{i=1}^{n-1} A_i\right)$$

$$\leq \sum_{i=1}^{n-1} P(A_i) + P(A_n) \leq \sum_{i=1}^{n} P(A_i).$$

$\square$

Let Ω be finite, $|\Omega| = N$. The probability distribution for which $p_i = 1/N$ is called uniform. Uniform distributions arise naturally in many applications. For example return to the problem of the "6 from 36" lottery. The space Ω in this case was constructed earlier, and consists of $\binom{36}{6}$ points. We now reason as follows. Let us fix a specific set $\bar{\omega}$. The result of the random drawing of the lottery will give a certain set ω of 6 numbers. It is natural to assume that all sets are equivalent, and that the probability of each ω is the same and equal to $p(\omega) = 1/N = 1/\binom{36}{6}$. We now find the probability of winning. According to the rules, winning will occur when ω and $\bar{\omega}$ have at least 3 numbers in common. Let us denote by C_k the event consisting of those ω such that ω and $\bar{\omega}$ have k numbers in common. Then $C = \bigcup_{k=3}^{6} C_k$ and consists of the winning ω's. It is clear that $C_i \cap C_j = \emptyset$ for $i \neq j$. Therefore $P(C) = \sum_{k=3}^{6} P(C_k)$. Furthermore, $P(C_k) = \sum_{\omega_i \in C_k} p_i = \sum_{\omega_i \in C_k} 1/N = 1/N|C_k| = |C_k|/|\Omega|$. It is easy to understand that $|C_k| = \binom{6}{k} \cdot \binom{30}{6-k}$. Therefore, for instance

$$P(C_3) = \frac{\binom{6}{3} \cdot \binom{30}{3}}{\binom{36}{6}} = \frac{\frac{6.5.4}{1.2.3} \cdot \frac{30.29.28}{1.2.3}}{\frac{36.35.34.33.32.31}{1.2.3.4.5.6}} \simeq 0.0412.$$

1.2 Expectation and Variance for Discrete Random Variables

We now return to the general situation. Assume, as previously, that the space Ω is discrete and consider a random variable $\xi = f(\omega)$. Its values form a countable set $X = \{x_i\}$ with $f(\omega) \in X$. Now consider the events $C_i = \{\omega : f(\omega) = x_i\}$. Clearly, $C_i \cap C_j = \emptyset$ for $i \neq j$ (the random variable takes only one value for each ω) and $\bigcup_{i=1}^{\infty} = \Omega$ (for each ω some value $f(\omega)$ is taken).

Definition 1.9. The collection of numbers $p_i = p(x_i) = P(C_i)$ defined on the set X is called the probability distribution of the random variable ξ.

The number p_i is the probability that the random variable ξ takes the value x_i. It follows from the properties of probabilities that $p_i \geq 0$, $\sum_i p_i = 1$.

Definition 1.10. The mathematical expectation (m.e.) or the mean value of the random variable ξ is the number

$$E\xi = \sum f(\omega)p(\omega),$$

which is defined when $\sum |f(\omega)|p(\omega) < \infty$. If the latter series diverges, the mathematical expectation is not defined.

In the case where the m.e. is defined, its value does not depend on the order of summation. The concept of m.e. is similar to the idea of center of gravity. In the same way as the center of gravity indicates near which point the mass is concentrated on average, the mathematical expectation indicates near which quantity the values of the random variable ξ are concentrated on average.

Properties of the Mathematical Expectation

1. If $E\xi_1$ and $E\xi_2$ are defined, then $E(a\xi_1 + b\xi_2)$ is defined, and $E(a\xi_1 + b\xi_2) = aE\xi_1 + bE\xi_2$.
2. If $\xi \geq 0$ then $E\xi \geq 0$.
3. If $\xi = f(\omega) \equiv 1$, then $E\xi = 1$.
4. If $A \leq \xi = f(\omega) \leq B$ then $A \leq E\xi \leq B$.
5. $E\xi$ is defined if and only if $\sum_i |x_i|p_i < \infty$ and then $E\xi = \sum_i x_i p_i$.
6. If the random variable $\eta = g(\xi) = g(f(\omega))$, then

$$E\eta = \sum g(x_i)p_i,$$

   where $E\eta$ is defined if and only if $\sum |g(x_i)|p_i < \infty$.
7. *Chebyshev's Inequality (1).* If $\xi \geq 0$ and $E\xi < \infty$, then for any $t > 0$

$$P\{(\omega| f(\omega) \geq t)\} \leq E\xi/t.$$

Proof of Properties 1–6. Property 1 follows from the fact that if $\xi_1 = f_1(\omega)$, $\xi_2 = f_2(\omega)$ and $E\xi_1$ and $E\xi_2$ are defined, then

$$\sum |f_1(\omega)|p(\omega) < \infty, \quad \sum |f_2(\omega)|p(\omega) < \infty.$$

Therefore

$$\sum |af_1(\omega) + bf_2(\omega)|p(\omega) \leq \sum (|a| |f_1(\omega)| + |b| |f_2(\omega)|)p(\omega)$$

$$= |a| \sum |f_1(\omega)|p(\omega) + |b| \sum |f_2(\omega)|p(\omega) < \infty$$

and, using the properties of absolutely converging series we find that

$$\sum (af_1(\omega) + bf_2(\omega))p(\omega) = a \sum f_1(\omega)p(\omega) + b \sum f_2(\omega)p(\omega).$$

Properties 2 and 3 are clear. Properties 1–3 mean that E is a linear (property 1) non-negative (property 2) normed (property 3) functional on the vector space of random variables. Property 4 follows from the fact that $\xi - A \geq 0$, $B - \xi \geq 0$ and therefore $E(\xi - A) = E\xi - EA.1 = E\xi - A \geq 0$ and, analogously, $B - E\xi \geq 0$.

We now prove property 6, since 5 follows from 6 for $g(x) = x$. First let $\sum |g(x_i)|p_i < \infty$. Since the sum of a series with non-negative terms does not depend on the order of the terms, $\sum_\omega |g(f(\omega))|p(\omega)$ can be carried out in the following way:

$$\sum_\omega |g(f(\omega))|p(\omega) = \sum_i \sum_{\omega \in C_i} |g(f(\omega))|p(\omega)$$

$$= \sum_i |g(x_i)| \sum_{\omega \in C_i} p(\omega) = \sum_i |g(x_i)|p_i.$$

Thus the series $\sum |g(f(\omega))|p(\omega)$ converges if and only if the series $\sum |g(x_i)|p_i$ does. If any of those series converges then the series $\sum g(f(\omega))p(\omega)$ converges absolutely, and its sum does not depend on the order of summation. Therefore

$$\sum_\omega g(f(\omega))p(\omega) = \sum_i \sum_{\omega \in C_i} g(f(\omega))p(\omega)$$

$$= \sum_i g(x_i) \sum_{\omega \in C_i} p(\omega) = \sum_i g(x_i)p_i,$$

and the last series also converges absolutely. □

Proof of Property 7. Let

$$P\{(\omega|f(\omega) \geq t)\} = \sum_{\omega|f(\omega) \geq t} p(\omega) \leq \sum_{\omega|f(\omega) \geq t} (f(\omega)/t)p(\omega)$$

$$= \frac{1}{t} \sum_{\omega|f(\omega) \geq t} f(\omega)p(\omega) \leq \frac{1}{t} \sum_\omega f(\omega)p(\omega) = \frac{1}{t}E\xi.$$

In the last inequality we used the fact that $f \geq 0$. □

In subsequent lectures we will encounter many problems in which the mathematical expectation plays a very important role. However we now introduce another numerical characteristic of a random variable.

Definition 1.11. The variance of a random variable ξ is the quantity $\mathrm{Var}\,\xi = E(\xi - E\xi)^2$.

The definition is understood to imply that $E\xi$ is defined. $\mathrm{Var}\,\xi$ characterizes the amount of variation of the random variable from its mean.

Properties of the Variance

1. $\mathrm{Var}\,\xi$ is finite if and only if $E\xi^2 < \infty$ and $\mathrm{Var}\,\xi = E\xi^2 - (E\xi)^2$.

2. $\mathrm{Var}(a\xi + b) = a^2\,\mathrm{Var}\xi$, where a and b are constants.
3. If $A \leq \xi \leq B$, then $\mathrm{Var}\xi \leq ((B - A)/2)^2$.
4. *Chebyshev's inequality (2).* Let $\mathrm{Var}\xi < \infty$. Then

$$P\{\omega|\,|f(\omega) - E\xi| \geq t\} \leq \frac{\mathrm{Var}\xi}{t^2}.$$

Proof of Property 1. Let $E\xi$ be defined, $|E\xi| < \infty$. Assume at first that $E\xi^2 < \infty$. Then $(\xi - E\xi)^2 = \xi^2 - 2(E\xi)\xi + (E\xi)^2$ and from property 1 of the m.e. $\mathrm{Var}\xi = E\xi^2 - E\big(2(E\xi)\xi\big) + E\big((E\xi)^2\big) = E\xi^2 - 2(E\xi)(E\xi) + (E\xi)^2 = E\xi^2 - (E\xi)^2 < \infty$. Now let $\mathrm{Var}\xi < \infty$. We have $\xi^2 = (\xi - E\xi)^2 + 2(E\xi)\xi - (E\xi)^2$, and from property 1 of the m.e.. We have

$$E\xi^2 = E(\xi - E\xi)^2 + 2(E\xi)^2 - (E\xi)^2 = \mathrm{Var}\xi + (E\xi)^2.$$

$\square$

Proof of Property 2. From property 1 of the m.e. we have $E(a\xi + b) = aE\xi + b$. Therefore

$$\mathrm{Var}(a\xi + b) = E(a\xi + b - E(a\xi + b))^2 = E(a\xi - aE\xi)^2$$
$$= E\big(a^2(\xi - E\xi)^2\big) = a^2\,E(\xi - E\xi)^2 = a^2\,\mathrm{Var}\xi.$$

$\square$

Proof of Property 3. Let $A \leq \xi \leq B$. It then follows from property 2 of the variance that

$$\mathrm{Var}\xi = E(\xi - E\xi)^2 = E(\xi - (A + B)/2 - (E\xi - (A + B)/2))^2$$
$$= E(\xi - (A + B)/2)^2 - \big(E(\xi - (A + B)/2)\big)^2 \leq E(\xi - (A + B)/2)^2.$$

We now note that $|\xi - (A+B)/2| \leq (B-A)/2$, $(\xi - (A+B)/2)^2 \leq ((B-A)/2)^2$. From property 4 of the m.e. we have

$$E(\xi - (A + B)/2)^2 \leq ((B - A)/2)^2.$$

$\square$

Proof of Property 4. We use Chebyshev's inequality (1) for the random variable $\eta = (\xi - E\xi)^2 \geq 0$. Then

$$\{\omega|\,|\xi - E\xi| \geq t\} = \{\omega|\,\eta \geq t^2\}.$$

Therefore

$$P\{\omega|\,|f(\omega) - E\xi| \geq t\} = P\{\omega|\,\eta(\omega) \geq t^2\} \leq \frac{E\eta}{t^2} = \frac{\mathrm{Var}\xi}{t^2}.$$

$\square$

Now let ξ_1, ξ_2 be two random variables taking the values $x_1, x_2, \ldots$ and $y_1, y_2, \ldots$ respectively.

Definition 1.12. The joint distribution of the random variables ξ_1, ξ_2 is the collection of numbers $p_{ij} = p(x_i, y_j) = P\{\omega | \, \xi_1(\omega) = x_i, \, \xi_2(\omega) = y_j\}$.

Given the joint distribution of ξ_1, ξ_2 one can find the distribution of ξ_1 (ξ_2) by summing the p_{ij} with respect to j (i). Moreover, if $\eta = f(\xi_1, \xi_2)$ then $E\eta = \sum f(x_i, y_j) p_{ij}$, again under the assumption that the last series converges absolutely. In particular $E\xi_1 . \xi_2 = \sum x_i y_j p_{ij}$.

One can define, in an analogous way, the joint distribution of any finite number of random variables.

Definition 1.13. The covariance coefficient, or simply the covariance of the random variables ξ_1, ξ_2 is the number $\mathrm{Cov}(\xi_1, \xi_2) = E(\xi_1 - m_1)(\xi_2 - m_2)$, where $m_i = E\xi_i$ for $i = 1, 2$.

We note that

$$\mathrm{Cov}(\xi_1, \xi_2) = E(\xi_1 - m_1)(\xi_2 - m_2) = E(\xi_1 \xi_2 - m_1 \xi_2 - m_2 \xi_1 + m_1 m_2)$$
$$= E\xi_1 \xi_2 - m_1 m_2.$$

Let $\xi_1, \xi_2, \ldots, \xi_n$ be random variables and $\varsigma_n = \xi_1 + \xi_2 + \cdots + \xi_n$. Then if $m_i = E\xi_i$, $E\varsigma_n = \sum_{i=1}^{n} m_i$,

$$\mathrm{Var}\varsigma_n = E\left(\sum_{i=1}^{n} \xi_i - \sum_{i=1}^{n} m_i\right)^2 = E\left(\sum_{i=1}^{n}(\xi_i - m_i)\right)^2$$

$$= \sum_{i=1}^{n} E(\xi_i - m_i)^2 + 2\sum_{i<j} E(\xi_i - m_i)(\xi_j - m_j)$$

$$= \sum_{i=1}^{n} \mathrm{Var}\xi_i + 2\sum_{i<j} \mathrm{Cov}(\xi_i, \xi_j).$$

Definition 1.14. The correlation coefficient of the random variables ξ_1, ξ_2 is the number $\rho(\xi_1, \xi_2) = \mathrm{Cov}(\xi_1, \xi_2) / \sqrt{\mathrm{Var}\xi_1 . \mathrm{Var}\xi_2}$.

Theorem 1.6. $|\rho(\xi_1, \xi_2)| \le 1$. If $|\rho(\xi_1, \xi_2)| = 1$ then $\xi_2 = a\xi_1 + b$, where a and b are constants.

Proof. The statement of the theorem is a simple consequence of the Cauchy-Bunyakovskii inequality. For every t we have

$$E\big(t(\xi_2 - m_2) + (\xi_1 - m_1)\big)^2 \ge 0.$$

Furthermore

$$E\big(t(\xi_2 - m_2) + (\xi_1 - m_1)\big)^2$$
$$= t^2 E(\xi_2 - m_2)^2 + 2t E(\xi_1 - m_1)(\xi_2 - m_2) + E(\xi_1 - m_1)^2$$
$$= t^2 \mathrm{Var}\xi_2 + 2t \mathrm{Cov}(\xi_1, \xi_2) + \mathrm{Var}\xi_1.$$

Since the latter quadratic polynomial (in t) is non-negative,

$$\big(\mathrm{Cov}(\xi_1, \xi_2)\big)^2 \leq \mathrm{Var}\xi_1 . \mathrm{Var}\xi_2,$$

i.e. $|\rho(\xi_1, \xi_2)| \leq 1$. If $|\rho(\xi_1, \xi_2)| = 1$ then for some t_0 we have $E\big(t_0(\xi_2 - m_2) + (\xi_1 - m_1)\big)^2 = 0$, i.e. $t_0(\xi_2 - m_2) + (\xi_1 - m_1) = 0$ and $\xi_2 = m_2 - (m_1/t_0) + (\xi_1/t_0)$. Setting $a = 1/t_0$, $b = m_2 - (m_1/t_0)$ we obtain the statement of the theorem. $\qquad\square$

Lecture 2. Independent Identical Trials and the Law of Large Numbers

2.1 Algebras and σ-algebras; Borel σ-algebras

We now consider some of the most important examples of σ-algebras encountered in probability theory. First we introduce the following general definition.

Definition 2.1. Let $\mathcal{A} = \{A\}$ be an arbitrary family of subsets of Ω. The smallest σ-algebra which contains $\mathcal{A}$ is called the σ-algebra generated by $\mathcal{A}$. This σ-algebra will be denoted by $\mathcal{F}(\mathcal{A})$.

In other words we can say that $\mathcal{F}(\mathcal{A})$ is the intersection of all those σ-algebras G which contain $\mathcal{A}$, $(\mathcal{A} \subset G)$. There is at least one such σ-algebra, namely the σ-algebra of all subsets of Ω. A less formal definition of $\mathcal{F}(\mathcal{A})$ consists in the following. Given sets from $\mathcal{A}$ we form finite and infinite unions of those sets. From the obtained subsets we form finite and infinite intersections. Then once again we form unions and intersections and so forth. The complete process makes use of the concept of transfinite induction, but we will not elaborate on this point.

Assume now that $\Omega = \mathbb{R}^1$. Consider the following families of subsets:

1. $\mathcal{A}_1$ is the collection of intervals $A \in \mathcal{A}_1$ which have the form $A = (a, b)$;
2. $\mathcal{A}_2$ is the collection of half-open intervals $A = [a, b)$;
3. $\mathcal{A}_3$ is the collection of half-open intervals $A = (a, b]$;
4. $\mathcal{A}_4$ is the collection of closed intervals $A = [a, b]$;
5. $\mathcal{A}_5$ is the collection of left intervals $(-\infty, a)$;
6. $\mathcal{A}_6$ is the collection of right intervals (a, ∞);
7. $\mathcal{A}_7$ is the collection of open sets of $\mathbb{R}^1$;
8. $\mathcal{A}_8$ is the collection of closed subsets of $\mathbb{R}^1$.

Theorem 2.1. $\mathcal{F}(\mathcal{A}_1) = \mathcal{F}(\mathcal{A}_2) = \ldots = \mathcal{F}(\mathcal{A}_8) = \mathcal{F}(\mathbb{R}^1)$. *The σ-algebra $\mathcal{F}(\mathbb{R}^1)$ is called the Borel σ-algebra or the σ-algebra of Borel subsets.*

Proof. All the equalities are proved in the same way. Let us prove for example that $\mathcal{F}(\mathcal{A}_5) = \mathcal{F}(\mathcal{A}_1)$. First for any $a < b$ we have the equality $(-\infty, b) \setminus (-\infty, a) = [a, b) \in \mathcal{F}(\mathcal{A}_5)$. Let us further take an interval (a, b) and a sequence $a_n \downarrow a$. Then $\bigcup_n [a_n, b) = (a, b)$ and $\bigcup_n [a_n, b) \in \mathcal{F}(\mathcal{A}_5)$. Therefore $(a, b) \in$

$\mathcal{F}(\mathcal{A}_5)$, i.e. $\mathcal{F}(\mathcal{A}_1) \subseteq \mathcal{F}(\mathcal{A}_5)$. One can prove in an analogous way that $\mathcal{F}(\mathcal{A}_1) \supseteq \mathcal{F}(\mathcal{A}_5)$. $\qquad\qquad\qquad\qquad\qquad\qquad\qquad\qquad\qquad\qquad\qquad\qquad\qquad\qquad\square$

Let us assume that Ω is a metric space.

Definition 2.2. The Borel σ-algebra on Ω is the σ-algebra $\mathcal{F}(\mathcal{A})$, where $\mathcal{A}$ is the family of open sets of Ω.

Now let Ω be the space of words $\omega = \{x_1, \ldots, x_n\}$ where $x_i \in X$, i.e. $\Omega = X^{[1,n]}$. We recall that X is an "alphabet". We assume that we have fixed on X a σ-algebra of subsets $\mathcal{B}$.

Definition 2.3. A finite-dimensional cylinder C is a set of the form

$$C = \{\omega | x_{i_1} \in B_1, \, x_{i_2} \in B_2, \ldots, x_{i_r} \in B_r\},$$

where $i_1, \ldots, i_r$ are given indices and $B_1, B_2, \ldots, B_r \in \mathcal{B}$.

In what follows, when we talk of a σ-algebra of subsets of $\Omega = X^{[1,n]}$ we will mean the σ-algebra $\mathcal{F} = \mathcal{F}(\{C\})$ where $\{C\}$ is the family of cylinders C.

This definition can be extended, without any changes, to spaces Ω of the form $\Omega = X^T$.

2.2 Heuristic Approach to the Construction of "Continuous" Random Variables

Earlier we introduced the concept of a probability distribution for a random variable $\xi = f(\omega)$, defined on the discrete space Ω. Such a random variable takes a finite or countable number of values $x_1, x_2, \ldots$.

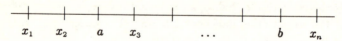

If the probability of occurrence of the value x_i is equal to p_i then we consider that the point x_i carries the mass p_i. It follows that for any interval $[a, b]$ the probability $P\{a \leq \xi \leq b\} = \sum_{x_i | a \leq x_i \leq b} p_i$, i.e. the probability that the random variable takes values in the interval $[a, b]$ is equal to the sum of the masses p_i which lie in the interval. Imagine that the number of values taken by the random variable converges to infinity and that the distances $x_{i+1} - x_i$ as well as the p_i converge to zero, where $\sum_{a \leq x_i \leq b} p_i$ converges to a finite value. For example let $-R \leq x_i \leq R$, $x_i = i/N$ and $p_i = p(\frac{i}{N})\frac{1}{N} + o(\frac{1}{N})$, where $p(t)$ is a function on the real line. Then the sum

$$\sum_{a \leq x_i \leq b} p_i = \sum_{a \leq i/N \leq b} \frac{1}{N} p\left(\frac{i}{N}\right) + o\left(\frac{1}{N}\right) N$$

behaves like a Riemann sum for the integral $\int_a^b p(t)dt$. It is clear that $p(t) \geq 0$ and $\int_{-\infty}^{\infty} p(t)dt = 1$, since $\sum_i p_i = \sum_i \frac{1}{N} p(\frac{i}{N}) + O(\frac{1}{N}) = 1$. Any function $p(t) \geq 0$ such that $\int_{-\infty}^{\infty} p(t)dt = 1$ is called a probability density.

Definition 2.4. Let $(\Omega, \mathcal{F}, P)$ be a probability space, and $\xi = f(\omega)$ be a function on Ω. If for any interval $[a, b]$

$$P\{a \leq \xi \leq b\} = P\{\omega | a \leq f(\omega) \leq b\} = \int_b^b p(t)dt$$

we say that the random variable ξ has a probability distribution with density p.

Examples of Densities

1. $p(x) = \frac{1}{\sqrt{2\pi}} e^{-x^2/2}$, $-\infty < x < \infty$ is called the normal density or the Gaussian distribution (with parameters $(0, 1)$);

1'. $p(x) = \frac{1}{\sqrt{2\pi}\sigma} \exp(-\frac{(x-m)^2}{2\sigma})$ is the normal density with parameters (m, σ);

2.

$$p(x) = \begin{cases} \frac{1}{b-a}, & x \in [a, b] \\ 0, & x \notin [a, b] \end{cases}$$

is the uniform density on the interval $[a, b]$;

2'. The random variables which we have considered thus far take real values, i.e. values in $\mathbb{R}^1$. One also encounters more general cases (see below) when the random variables take values in the space $\mathbb{R}^d$. Let us fix a subset $V \subseteq \mathbb{R}^d$. The uniform distribution on V is the distribution of a random variable with values in V such that for any $U \subset V$

$$P\{\xi \in U\} = \frac{\text{vol } U}{\text{vol } V}.$$

We assume here that $0 < \text{vol } V < \infty$ (vol $(\cdot)$ represents the d-dimensional volume of a set). Sometimes to define the uniform distribution on V one says "a point is chosen in the set V at random with the uniform distribution";

3.

$$p(x) = \begin{cases} \lambda e^{-\lambda x}, & x \geq 0, \\ 0, & x < 0 \end{cases}$$

is the exponential density;

4. $p(x) = 1/\pi(1 + x^2)$, $-\infty < x < \infty$ is the Cauchy density.

The mathematical expectation of a random variable which has a probability density $p(x)$ is given by the integral $E\xi = \int_{-\infty}^{\infty} xp(x)\,dx$, and this mathematical expectation is defined if the latter integral converges absolutely. In exactly the same way

$$\text{Var}\xi = \int_{-\infty}^{\infty} (x - E\xi)^2 p(x)\,dx = \int_{-\infty}^{\infty} x^2 p(x)\,dx - \left(\int_{-\infty}^{\infty} xp(x)\,dx\right)^2.$$

Given two random variables ξ_1, ξ_2 their joint probability density is a function $p(x,y) \geq 0$ such that $\int_{-\infty}^{\infty} \int_{-\infty}^{\infty} p(x,y)\,dx\,dy = 1$ and such that $P\{a_1 \leq \xi_1 \leq b_1, a_2 \leq \xi_2 \leq b_2\} = \int_{a_1}^{b_1} \int_{a_2}^{b_2} p(x,y)\,dx\,dy$ for any $a_1 < b_1$, $a_2 < b_2$. It follows that ξ_1 (ξ_2) has a distribution with density $\int_{-\infty}^{\infty} p(x,y)\,dy (\int_{-\infty}^{\infty} p(x,y)\,dx)$. Furthermore

$$E\xi_1 \cdot \xi_2 = \int_{-\infty}^{\infty} \int_{-\infty}^{\infty} xy\,p(x,y)\,dx\,dy$$

assuming that the latter integral converges absolutely, and

$$\mathrm{Cov}(\xi_1, \xi_2) = \int_{-\infty}^{\infty} \int_{-\infty}^{\infty} xy\,p(x,y)\,dx\,dy$$
$$- \int_{-\infty}^{\infty} \int_{-\infty}^{\infty} x\,p(x,y)\,dx\,dy \cdot \int_{-\infty}^{\infty} \int_{-\infty}^{\infty} y\,p(x,y)\,dx\,dy.$$

2.3 Sequences of Independent Trials

Let $X = \{x^{(1)}, x^{(2)}, \ldots, x^{(r)}\}$ be a finite set on which a probability distribution $p = \{p_1, \ldots, p_r\}$, i.e. a collection of $p_i \geq 0$ with $\sum_{i=1}^{r} p_i = 1$, is given. The individual number p_i is called the probability of the outcome $x^{(i)}$. We will also consider the p_i as a function given on X, $p_i = p(x^{(i)})$. Consider the space of elementary outcomes $\Omega = X^{[1,n]}$, i.e. the space of words $\omega = \{x_1, \ldots, x_n\}$, where $x_k \in X$. We set $p(\omega) = \prod_{i=1}^{n} p(x_i)$. Clearly $p(\omega) \geq 0$. We show that $\sum_{\omega} p(\omega) = 1$. For $n = 1$ this follows from the fact that $\sum p(\omega) = \sum_{x_1 \in X} p(x_1) = \sum_{k=1}^{r} p(x^{(k)}) = 1$. We now argue by induction. The summation $\sum_n = \sum_{\{x_1, \ldots, x_n\}} p(\omega)$ can be carried out in the following way: first fix $x_1, \ldots, x_{n-1}$ and sum over all values of x_n, and then sum the result over $x_1, \ldots, x_{n-1}$:

$$\sum_n = \sum_{x_1, \ldots, x_{n-1}} \sum_{x_n} p(x_1) \ldots p(x_{n-1}) p(x_n) = \sum_{x_1, \ldots, x_{n-1}} p(x_1) \ldots p(x_{n-1}).$$

The last sum is equal to 1, since $\sum_{x_n \in X} p(x_n) = \sum_{k=1}^{r} p(x^{(k)}) = 1$. So $\sum_n = \sum_{n-1}$, and by the induction hypothesis $\sum_n = 1$. It follows that the numbers $p(\omega)$ define a probability distribution P on Ω.

Definition 2.5. The probability space $(\Omega, \mathcal{F}, P)$ is called a sequence of n independent identical trials with r outcomes and with distribution of outcomes P. If $|X| = 2$ then a sequence of n independent identical trials is called a sequence of identical Bernoulli trials. If $X = \{1, 0\}$ or $X = \{1, -1\}$ then sometimes 1 is called the "success" or "win" and 0 or -1 are called "failure" or "loss".

We now study the properties of P. Let us choose indices $i_1, i_2, \ldots, i_t$ and as many outcomes $x^{(j_1)}, x^{(j_2)}, \ldots, x^{(j_t)}$ (among which there may be repetitions) and consider the cylinder

$$C = \{(\omega = (x_1, \ldots, x_n) | x_{i_1} = x^{(j_1)}, \ldots, x_{i_t} = x^{(j_t)})\}.$$

Lemma 2.1. $P(C) = p(x^{(j_1)}) \cdots p(x^{(j_t)}).$

Proof. $P(C) = \sum p(\omega) = \sum p(x_1) \ldots p(x_n)$ where the summation is carried out over those $\omega = (x_1, \ldots, x_n)$ for which $p(x_{i_\ell}) = p(x^{(j_\ell)})$, $1 \le \ell \le t$. Then

$$P(C) = \sum_{\substack{x_i, i \ne i_1, \ldots, i_t \\ x_{i_\ell} = x^{(j_\ell)}, 1 \le \ell \le t}} p(x_1) \ldots p(x_n)$$

$$= p(x^{(j_1)}) \ldots p(x^{(j_t)}) \sum_{x_i, i \ne i_1, \ldots, i_t} \prod_{i \ne i_1, \ldots, i_t} p(x_i).$$

The last sum is equal to 1. This is proved in the same way as the normalization property $\sum p(\omega) = 1$ was proved before.   □

Corollary 2.1. *Let* $C_i \subset X$, $1 \le i \le n$ *be arbitrary subsets of* X *and let* $C = \{\omega | \omega_i \in C_i, 1 \le i \le n\}$. *Then* $P(C) = \prod_{i=1}^n P(C_i)$.

Proof. We have

$$P(C) = \sum_{\omega \in C} p(\omega) = \sum_{\substack{\omega | x_i \in C_i \\ 1 \le i \le n}} p(x_1) \ldots p(x_n)$$

$$= \prod_{i=1}^n \sum_{x_i \in C_i} p(x_i) = \prod_{i=1}^n P(C_i).$$

□

We introduce the random variable $\nu^{(j)}(\omega)$, equal to the number of i, for which $1 \le i \le n$ and $x_i = x^{(j)}$, and the random variables $\xi_i^{(j)}(\omega)$ where

$$\xi_i^{(j)}(\omega) = \begin{cases} 1, & \text{if } x_i = x^{(j)}, \\ 0, & \text{otherwise.} \end{cases}$$

Then

$$\nu^{(j)}(\omega) = \sum_{i=1}^n \xi_i^{(j)}(\omega) \qquad (2.1)$$

It is clear that $\nu^{(j)}(\omega)$ takes the values $0, 1, \ldots, n$.

Theorem 2.2. $P\{\nu^{(j)} = k\} = \binom{n}{k}(p^{(j)})^k(1 - p^{(j)})^{n-k}$, $E\nu^{(j)} = np^{(j)}$, $\mathrm{Var}\nu^{(j)} = np^{(j)}(1 - p^{(j)})$.

Proof. We note that $1 - p^{(j)} = \sum_{s \ne j} p^{(s)}$. Let us define $B_k = \{\omega | \nu^{(j)}(\omega) = k\}$. Then

$$B_k = \bigcup B_{k; i_1, \ldots, i_k},$$

where $1 \leq i_1 < i_2 < \ldots < i_k = n$ and $x_{i_\ell} = x^{(j)}$, $1 \leq \ell \leq k$, $x_i \neq x^{(j)}$ for the remaining i. The number of events $B_{k;i_1,\ldots,i_k}$ is clearly equal to $\binom{n}{k}$, and they do not intersect. Furthermore

$$P(B_{k;i_1,\ldots,i_k}) = \sum_{\omega \in B_{k;i_1,\ldots,i_k}} p(\omega)$$

$$= \sum_{\omega \in B_{k;i_1,\ldots,i_k}} p(x_1)p(x_2)\ldots p(x_k)$$

$$= (p^{(j)})^k \sum_{\substack{x_i \neq x^{(j)} \\ i \neq i_1,\ldots,i_k}} \prod_{i \neq i_1,\ldots,i_k} p(x_i).$$

In the last sum we can interchange product and summation:

$$\sum_{\substack{x_i \neq x^{(j)} \\ i \neq i_1,\ldots,i_k}} \prod_{i \neq i_1,\ldots,i_k} p(x_i) = \prod_{i \neq i_1,\ldots,i_k} \sum_{x_i \neq x(j)} p(x_i) = (1 - p^{(j)})^{n-k}.$$

In the same way

$$P(B_{k;i_1,\ldots,i_k}) = (p(j))^k (1 - p^{(j)})^{n-k},$$

$$P(B_k) = \binom{n}{k}(p^{(j)})^k (1 - p^{(j)})^{n-k}.$$

In order to calculate $E\nu^{(j)}$ and $\mathrm{Var}\nu^{(j)}$ we use (2.1) and Lemma 2.1 where, for C, we take the cylinder $C = \{\omega | x_{i_1} = x^{(j)}, x_{i_2} = x^{(j)}\}$:

$$E\nu^{(j)} = \sum_{i=1}^{n} E\xi_i^{(j)} = \sum_{i=1}^{n} P(\xi_i^{(j)} = 1) = np^{(j)},$$

$$\mathrm{Var}\nu^{(j)} = E(\nu^{(j)})^2 - (E\nu^{(j)})^2$$

$$= E\left(\sum_{i=1}^{n} \xi_i^{(j)}\right)^2 - n^2 (p^{(j)})^2$$

$$= \sum_{i_1,i_2} E\xi_{i_1}^{(j)} \xi_{i_2}^{(j)} - n^2 (p^{(j)})^2$$

$$= \sum_{i_1 = i_2} E(\xi_{i_1}^{(j)})^2 + \sum_{i_1 \neq i_2} E\xi_{i_1}^{(j)} \xi_{i_2}^{(j)} - n^2 (p^{(j)})^2$$

$$= np^{(j)} + n(n-1)(p^{(j)})^2 - n^2 (p^{(j)})^2$$

$$= np^{(j)} (1 - p^{(j)}).$$

Here we use the fact that $(\xi_{i_1}^{(j)})^2 = \xi_{i_1}^{(j)}$ and that, by Lemma 2.1,

$$E\xi_{i_1}^{(j)} \xi_{i_2}^{(j)} = P\{(\xi_{i_1}^{(j)} = 1, \xi_{i_2}^{(j)} = 1)\}$$

$$= P\{(\omega | x_{i_1} = x^{(j)}, x_{i_2} = x^{(j)})\} = (p^{(j)})^2.$$

□

We now choose two numbers $p, q \geq 0$ s.t. $p + q = 1$.

Definition 2.6. The probability distribution of the random variable ξ, which takes integer values $0 \leq k \leq n$, where $P\{\xi(\omega) = k\} = \binom{n}{k} p^k q^{n-k} = \binom{n}{k} p^k (1 - p)^{n-k}$ is called a binomial distribution with parameters $p, q = 1 - p$.

Theorem 2.2 shows that the random variable $\nu^{(j)}$ has a binomial distribution with parameters $p^{(j)}, 1 - p^{(j)}$.

From Theorem 2.2 we now obtain a very important corollary.

2.4 Law of Large Numbers for Sequences of Independent Identical Trials

For any $\delta > 0$

$$P\{(\omega| \, |\frac{\nu^{(j)}}{n} - p^{(j)}| \geq \delta \quad \text{for at least one } j, \quad 1 \leq j \leq r)\} \to 0 \quad \text{as} \quad n \to \infty.$$

This statement follows in a straightforward manner from the Chebyshev inequality. Let $A^{(j)} = (\omega| \, |\frac{\nu^{(j)}}{n} - p^{(j)}| \geq \delta)$. We are interested in $P(\bigcup_{j=1}^{r} A^{(j)})$. We have $P(\bigcup_{j=1}^{r} A^{(j)}) \leq \sum_{j=1}^{r} P(A^{(j)})$. Furthermore, by Chebyshev's inequality

$$P(A^{(j)}) = P\{(\omega| \, |\frac{\nu^{(j)}}{n} - p^{(j)}| \geq \delta)\}$$
$$= P\{(\omega| \, |\nu^{(j)} - np^{(j)}| \geq \delta n)\}$$
$$= P\{(\omega| \, |\nu^{(j)} - E\nu^{(j)}| \geq \delta n)\}$$
$$\leq \frac{\text{Var}\nu^{(j)}}{\delta^2 n^2} = \frac{np^{(j)}(1 - p^{(j)})}{\delta^2 n^2}$$
$$= \frac{p^{(j)}(1 - p^{(j)})}{\delta^2 n} \to 0 \quad \text{as} \quad n \to \infty.$$

Consequently $\sum_{j=1}^{r} P(A^{(j)}) \to 0$ as $n \to \infty$. We will now derive corollaries from the Law of Large Numbers (LLN). We note that the LLN gives a foundation for representing probabilities as limits of frequencies. This can be done if further information is available which allows us to assume that we are dealing with a probability distribution which corresponds to a sequence of independent trials. The LLN is a particular case of a general ergodic theorem on the behavior of arithmetic means of random variables, which holds under much more general assumptions.

2.5 Generalizations

The first generalization consists in the fact that we can consider an arbitrary probability space $(X, \mathcal{B}, \bar{P})$ instead of a finite X, where $\bar{P}$ is a probability defined on the events $B \in \mathcal{B}$. As previously, we set $\Omega = X^{[1,n]}$ and consider the cylinders $C = (\omega | x_{i_1} \in B_1, \ldots, x_{i_k} \in B_k)$, where $B_1, \ldots, B_k \in \mathcal{B}$.

Let $P(C) = \prod_{\ell=1}^{k} \bar{P}(B_\ell)$ for any cylinder C. One can prove (see below) that there exists a unique σ−additive set function defined on $\mathcal{B}$ which takes this form on cylinders C.

Definition 2.7. The probability space $(\Omega, \mathcal{F}, P)$ is called a sequence of n independent identical trials with space of outcomes X and probability distribution $\bar{P}$ for those outcomes.

The second generalization, which we mention only briefly, consists in considering spaces X which depend on the number of trials. We are then talking about the concept of sequences of independent non-identical trials.

2.6 Applications

2.6.1 Probabilistic Proof of Weierstrass's Theorem

Weierstrass's Theorem (WT), from courses in mathematical analysis, states that any continuous function $f(x)$, $x \in [0, 1]$ can be uniformly approximated by a polynomial. We give a probabilistic proof of the WT due to S. N. Bernstein.

Definition 2.8. The Bernstein polynomial of degree n for the function f is the polynomial $Q_n(x) = \sum_{k=0}^{n} \binom{n}{k} x^k (1-x)^{n-k} f(k/n)$.

Theorem 2.3 (Weierstrass). *For any continuous function f*

$$\sup_{x \in [0,1]} |Q_n(x) - f(x)| \to 0 \qquad \text{as} \quad n \to \infty.$$

Proof. Let $\epsilon > 0$. Since the function f is continuous on $[0, 1]$ it is uniformly continuous. Therefore there exists a $\delta > 0$ such that the inequality $|x' - x''| \leq \delta$ implies $|f(x') - f(x'')| \leq \epsilon/2$. We define $M = \sup_{x \in [0,1]} |f(x)|$.

For any n we have

$$f(x) - Q_n(x) = f(x) - \sum_{k=0}^{n} \binom{n}{k} x^k (1-x)^{n-k} f(k/n)$$

$$= \sum_{k=0}^{n} \binom{n}{k} x^k (1-x)^{n-k} (f(k/n) - f(x))$$

$$= \sum_{k:|k/n-x|<\delta} \binom{n}{k} x^k (1-x)^{n-k} (f(k/n) - f(x))$$

$$+ \sum_{k:|k/n-x|\geq\delta} \binom{n}{k} x^k (1-x)^{n-k} (f(k/n) - f(x))$$

$$= \Sigma_1 + \Sigma_2$$

The first sum can be estimated as follows:

$$|\Sigma_1| \leq \sum_{k:|k/n-x|<\delta} \binom{n}{k} x^k (1-x)^{n-k} |f(k/n) - f(x)|$$

$$\leq \frac{\epsilon}{2} \cdot \sum_{k:|k/n-x|<\delta} \binom{n}{k} x^k (1-x)^{n-k} \leq \frac{\epsilon}{2},$$

since the $\binom{n}{k} x^k (1-x)^{n-k}$ are the probabilities of the binomial distribution, and their sum over all k is equal to 1. As for the second sum, we have

$$|\Sigma_2| \leq \sum_{k:|k/n-x|\geq\delta} \binom{n}{k} x^k (1-x)^{n-k} |f(k/n) - f(x)|$$

$$\leq 2M \sum_{k:|k/n-x|\geq\delta} \binom{n}{k} x^k (1-x)^{n-k}.$$

At this point the probabilistic part of the argument begins. We consider a sequence of n independent trials with $X = \{1, 0\}$ and probabilities $p^{(1)} = x$, $p^{(0)} = 1 - x$ and the random variable $\nu^{(1)}$, equal to the number of occurrences of 1 among the n trials. Then $P(\nu^{(1)} = k) = \binom{n}{k} x^k (1-x)^{n-k}$ (from Theorem 2.2) and $\sum_{k:|k/n-x|\geq\delta} \binom{n}{k} x^k (1-x)^{n-k}$ is exactly equal to the probability that $|\nu^{(1)}/n - x| \geq \delta$. We saw in the proof of the LLN that this probability, by Chebyshev's inequality, is not greater than $x(1-x)/\delta^2 n$. Therefore $|\Sigma_2| \leq 2Mx(1-x)/\delta^2 n$. For all sufficiently large values of n this last expression will be smaller than $\epsilon/2$. Consequently $|\Sigma_1 + \Sigma_2| \leq \epsilon$ for all sufficiently large n. $\square$

2.6.2 Application to Number Theory

We again consider the interval $[0, 1]$. Let us expand every $x \in [0, 1]$ into its decimal expansion $x = 0, a_1, a_2 \ldots a_n \ldots$, where each digit a_i takes its values from 0 to 9. Let us fix the n first digits of the decimal expansion $a_1 \ldots a_n$. Those numbers $x \in [0, 1]$ which have these digits in their n first positions form an interval $\Delta(a_1 \ldots a_n)$ of length 10^{-n}. For different choices of $(a_1, a_2 \ldots a_n)$ those intervals either do not intersect, or intersect at only one end point. We consider the sequence of n independent identical trials, with space of outcomes $(0, 1, 2, \ldots, 9)$ and probability of each outcome equal to 10^{-1}. Then $\Omega^{[1,n]}$ consists of the words $\omega = (a_1 \ldots a_n)$ and $p(\omega) = 10^{-n}$. We denote by $\nu^{(j)}(x)$ the number of occurrences of the digit j among the n first digits of the decimal expansion. Let us fix a number $\delta > 0$ and consider those intervals $\Delta(a_1 \ldots a_n)$ for which $|\nu^{(j)}/n - 1/10| \leq \delta$, $0 \leq j \leq 9$. It then follows from the LLN that the sum of the lengths of those intervals converges to 1 as $n \to \infty$. One sometimes says that, in the decimal expansion of typical numbers x, all digits are encountered approximately the same number of times.

2.6.3 Monte Carlo Methods

We assume that we must calculate the d−dimensional integral

$$I = \int_0^1 \int_0^1 \cdots \int_0^1 f(t_1, \ldots, t_d) dt_1 \ldots dt_d$$

of a continuous function of d variables. In applications d can take values of order 10 and above. The computation of the usual Riemann sums uses an enormous volume of operations, and consequently of computer time. So already at the origin of the development of computational techniques a probabilistic method of computing these integrals was conceived which uses much less computation, but sometimes leads to gross errors. A sketch of the method is as follows: let $y = (y_1, \ldots, y_n)$ and assume that n points $y^{(1)}, \ldots y^{(n)}$ are chosen independently with a uniform distribution in the d−dimensional unit cube. We form the average

$$\frac{1}{n} \sum_{s=1}^{n} f(y^{(s)}).$$

It turns out that with large probability this average is close to I. The value of d is irrelevant to the proof of this statement, so we will assume that $d = 1$, i.e. that

$$I = \int_0^1 f(y)\, dy.$$

First we explain the meaning of the words "n points are chosen independently in the interval $[0, 1]$", which we now denote by $y_1, \ldots, y_n$. This means that we consider a sequence of n identical trials in which $X = [0, 1]$, and in which the distribution of outcomes is the uniform distribution on $[0, 1]$. The sum $(1/n) \sum_{i=1}^{n} f(y_i)$ is a random variable defined on the space Ω of words $\omega = (y_1, \ldots, y_n)$. We now derive the following statement from the LLN:

$$P\{(\omega \mid |\frac{1}{n} \sum_{i=1}^{n} f(y_i) - I| \leq \epsilon)\} \to 1 \quad \text{as} \quad n \to \infty.$$

For the proof we divide the interval $[0,1]$ into r equal parts by the points $x^{(i)} = i/r$, $0 \le i \le r$. Given $\omega = (y_1,\ldots,y_n) \in X^{[1,n]}$, we construct $\omega' = (x_1,\ldots,x_n)$, where $x_k = x^{(i)}$ if $i/r \le y_k \le (i+1)/r$ (in the case of two possibilities we choose the left hand side point). We now find the largest r such that $|f(x_k) - f(y_k)| \le \epsilon/2$. This is possible since the function f is continuous on $[0,1]$ and therefore is uniformly continuous. It follows that

$$\left| \frac{1}{n} \sum_{s=1}^{n} f(y_s) - \frac{1}{n} \sum_{s=1}^{n} f(x_s) \right| \le \frac{1}{n} \sum_{s=1}^{n} |f(y_s) - f(x_s)| \le \frac{\epsilon}{3}.$$

We now introduce on the space Ω' of words ω' the probability distribution which assigns to each ω' a probability $p(\omega')$ equal to the probability of those $\omega \in \Omega$ which correspond to ω'. But it is clear then that if $\omega' = (x_1,\ldots,x_n)$ then

$$p(\omega') = P\{(\omega = (y_1,\ldots,y_n) | x_i \le y_i \le x_i + \frac{1}{r}, 1 \le i \le n)\} = \left(\frac{1}{r}\right)^n.$$

The last relation shows that we have on the space Ω' a probability distribution which corresponds to a sequence of n independent trials with space of outcomes $X' = \{0, 1/r, 2/r, \ldots, (r-1)/r\}$ and probability $1/r$ for each outcome. By the LLN $P\{(\omega' | |\nu^{(j)}/n - 1/r| \le \delta, 0 \le j \le r-1)\} \to 1$ as $n \to \infty$ for any fixed δ. Here $\nu^{(j)} = \nu^{(j)}(\omega')$ is the number of occurrences of the number j/r in the word ω'. We can therefore write

$$\frac{1}{n} \sum_{i=1}^{n} f(x_i) = \sum_{j=0}^{r-1} \frac{\nu^{(j)}}{n} f(x^{(j)})$$

$$= \sum_{j=0}^{r-1} \frac{1}{r} f\left(\frac{j}{r}\right) + \sum_{j=0}^{r-1} \left(\frac{\nu^{(j)}}{n} - \frac{1}{r}\right) f\left(\frac{j}{r}\right).$$

We choose r such that $|\frac{1}{r} \sum_{j=0}^{r-1} f(\frac{j}{r}) - I| \le \frac{\epsilon}{3}$, and we set $\delta = \epsilon/3rM$, where $M = \max|f(x)|$. Then

$$\left| \sum_{j=0}^{r-1} \left(\frac{\nu^{(j)}}{n} - \frac{1}{r}\right) f\left(\frac{j}{r}\right) \right| \le \sum_{j=0}^{r-1} \left| \frac{\nu^{(j)}}{r} - \frac{1}{r} \right| \cdot \left| f\left(\frac{j}{r}\right) \right| \le Mr\epsilon(3rM)^{-1} = \frac{\epsilon}{3}.$$

Finally we obtain that if r is chosen in such a way that all the requirements mentioned above are satisfied, and if $y = (y^{(1)},\ldots,y^{(n)})$ is such that $|\nu^{(j)}/n - 1/r| \le \delta$ for all $0 \le j \le r-1$ then $|(1/n)\sum_{s=1}^{n} f(y^{(s)}) - I| \le \epsilon$. It follows from the LLN that the probability of such y converges to 1 as $n \to \infty$.

2.6.4 Entropy of a Sequence of Independent Trials and Macmillan's Theorem

As before, let $\Omega = X^{[1,n]}$ and $X = \{x^{(1)},\ldots,x^{(r)}\}$, $p^{(j)} = p(x^{(j)})$, $1 \le j \le r$.

Definition 2.9. The entropy of a sequence of independent identical trials is the number $h = -\sum_{j=1}^{r} p^{(j)} \ln p^{(j)}$.

Here we assume that $0 \ln 0 = 0$. Clearly $h \geq 0$ and $h = 0$ only in the degenerate case where one of the probabilities is equal to 1, and the remaining ones equal to zero. Furthermore $\max_{\{p^{(j)}\}} h = \ln r$. Indeed if all $p^{(j)} > 0$ then $h = h(\{p^{(j)}\})$ is a differentiable function of r variables $p^{(j)}$, subject to the condition $\sum_{j=1}^{r} p^{(j)} = 1$. To find the maximum of this function we use Laplace multipliers and set $\mathcal{F}(p^{(1)}, \ldots, p^{(r)}) = -\sum_{j=1}^{r} p^{(j)} \ln p^{(j)} + \lambda(\sum_{j=1}^{r} p^{(j)} - 1)$. Then $0 = \frac{\partial \mathcal{F}}{\partial p^{(j)}} = -\ln p^{(j)} - 1 + \lambda$, i.e. $p^{(j)} = e^{-1+\lambda}$. So all the $p^{(j)}$ are equal, $p^{(j)} = 1/r$, and consequently, for such a distribution, $h = \ln r$. If some of the probabilities are equal to zero, then the same argument shows that $\max h$ for such $p^{(j)}$ is equal to $\ln r'$, where r' is the number of non-zero probabilities, and we see that $\ln r' < \ln r$. So $\max h = \ln r$ and it is attained for the uniform distribution on X. The meaning of entropy is made clear by Macmillan's Theorem, which we quote below.

Theorem 2.4 (Macmillan). *For any $\alpha > 0$, $\beta > 0$, there exists a number $n_0(\alpha, \beta)$ such that for all $n > n_0(\alpha, \beta)$ there exists in the space $\Omega = X^{[1,n]}$ an event $C_n \subset \Omega$ which satisfies the following properties:*

1. $P(C_n) \geq 1 - \alpha$;
2. *for every* $\omega \in C_n$, $e^{-n(h+\beta)} \leq p(\omega) \leq e^{-n(h-\beta)}$;
3. $e^{n(h-\beta)} \leq |C_n| \leq e^{n(h+\beta)}$.

Proof. Let $\delta > 0$, whose value we will specify later, and set

$$C_n = \{\omega | \, |\nu^{(j)}/n - p^{(j)}| \leq \delta, \qquad 1 \leq j \leq r\}.$$

It follows from the LLN that $P(C_n) \to 1$ as $n \to \infty$ and therefore for $n \geq n_0(\delta, \alpha)$ we have $P(C_n) \geq 1 - \alpha$.

Let us now prove property 2. For $\omega = \{x_1, \ldots, x_n\}$ we have

$$p(\omega) = \prod_{s=1}^{n} p(x_s) = \prod_{j=1}^{r} (p^{(j)})^{\nu^{(j)}(\omega)}$$

$$= \exp\left(\sum_{j=1}^{r} \nu^{(j)}(\omega) \ln p^{(j)}\right) = \exp\left(-n\left(-\sum_{j=1}^{r} \frac{\nu^{(j)}}{n} \ln p^{(j)}\right)\right)$$

$$= \exp\left(-n\left(-\sum_{j=1}^{r} p^{(j)} \ln p^{(j)} - \sum_{j=1}^{r} \left(\frac{\nu^{(j)}}{n} - p^{(j)}\right) \ln p^{(j)}\right)\right).$$

Since $\omega \in C_n$ then

$$\left|\sum_{j=1}^{r} \left(\frac{\nu^{(j)}}{n} - p^{(j)}\right) \ln p^{(j)}\right| \leq \delta \sum_{j=1}^{r} |\ln p^{(j)}|.$$

We choose δ such that $\delta \sum_{j=1}^{r} |\ln p^{(j)}| \leq \beta/2$. Then

$$\exp(-n(h+\beta/2)) \leq \exp\left(-n(-\sum p^{(j)} \ln p^{(j)} + \beta/2)\right) \leq p(\omega)$$
$$\leq \exp\left(-n(-\sum p^{(j)} \ln p^{(j)} - \beta/2)\right) = \exp\left(-n(h-\beta/2)\right).$$

This completes the proof of 2. To prove 3 we write

$$1 - \alpha \leq P(C_n) = \sum_{\omega \in C_n} p(\omega) \leq 1$$

and

$$e^{-n(h+\beta/2)}|C_n| \leq \sum_{\omega \in C_n} p(\omega) \leq e^{-n(h-\beta/2)}|C_n|.$$

It follows that for n large enough,

$$e^{n(h-\beta)} \leq (1-\alpha)e^{n(h-\beta/2)} \leq P(C_n)e^{n(h-\beta/2)}$$
$$\leq |C_n| \leq e^{n(h+\beta/2)}P(C_n) \leq e^{n(h+\beta)}.$$

$$\square$$

The meaning of Macmillan's Theorem (MT) is as follows: Clearly $|\Omega| = r^n$. MT shows that we can choose, in Ω, a subset C_n for which the number of points is between $e^{n(h\pm\beta)}$, and whose probability is arbitrarily close to 1. If h is much smaller than $\ln r$, then such a reduction of the space Ω does not change much from the point of view of probability theory, but significantly reduces the space itself. This property plays an important role in many problems in information theory and ergodic theory. In addition the MT shows that for any arbitrary non-uniform distribution $\{p^{(j)}\}$ the probabilities $p(\omega)$ for large n become equal in the limit, i.e. the distribution approaches the uniform distribution, although in a very weak sense.

2.6.5 Random Walks

Let $X = \{-1,1\}$, $p(1) = p$, $p(-1) = q$, and consider the sequence of independent identical trials with $\Omega = X^{[1,n]}$. Here $\omega = \{x_1, \ldots, x_n\}$ where each $x_i = \pm 1$. We set $y_0 = 0$, $y_k = \sum_{i=1}^{k} x_i$, $1 \leq k \leq n$ and we draw in the plane (t, y) the broken line $y(t)$ which passes through the points (k, y_k), $0 \leq k \leq n$.

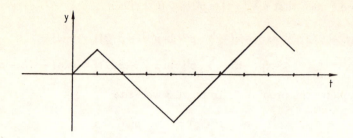

Fig. 2.1.

It is clear that the function $y(t)$ uniquely defines ω. Each $y(t)$ consists of straight segments, at angles of $\pm 45°$. If we consider $y(t)$ as the graph of a motion it is clear that for each unit of time the moving point moves along a segment of line 1, but arbitrarily changes the direction of motion. The function $y(t)$ is called the trajectory of the random walk. The trajectory of a random walk can be figuratively defined as the trajectory of a drunk person. It is clear that if k is even then $y(k)$ is even, and if k is odd $y(k)$ is odd. We set $P\{(y(t),\ 0 \le t \le n)\} = p(\omega)$, where $p(\omega)$ corresponds to the sequence $\omega = \{x_1,\ldots,x_n\}$, which gave $y(t)$. Then $P\{(y(t),\ 0 \le t \le n)\} = p^{\nu^{(1)}} q^{\nu^{(-1)}}$, where $\nu^{(1)} = \nu^{(1)}(\omega)$ is the number of unit segments which move according to an angle $+45°$, and $\nu^{(-1)} = \nu^{(-1)}(\omega)$ is the number of unit segments which move according to an angle of $-45°$. Clearly we have

$$y(n) = \nu^{(1)}(\omega) - \nu^{(-1)}(\omega), \quad \frac{y(n)}{n} = \frac{\nu^{(1)}(\omega)}{n} - \frac{\nu^{(-1)}(\omega)}{n}.$$

It follows from the LLN that for large n, with large probability, $n^{-1}\nu^{(1)}$ is close to p and $n^{-1}\nu^{(-1)}$ is close to q. Therefore $n^{-1}y(n)$ with large probability is close to $p - q$. The quantity $p - q$ is called the mean one-step displacement.

Definition 2.10. A random walk is said to be symmetric if $p - q = 0$, i.e. if $p = q = 1/2$.

We now find $Ey(n)$ and $Ey^2(n)$ for a symmetric random walk. We use the fact that $\nu^{(1)}(\omega) + \nu^{(-1)}(\omega) = n$, $\nu^{(-1)}(\omega) = n - \nu^{(1)}(\omega)$, $y(n) = 2\nu^{(1)}(\omega) - n$ and $\nu^{(1)}$ has a binomial distribution with $p = q = 1/2$. Using Theorem 2.1 we obtain:

$$Ey(n) = E(2\nu^{(1)}(\omega) - n) = 2E\nu^{(1)}(\omega) - n = 2.(n/2) - n = 0,$$

$$\mathrm{Var}\, y(n) = Ey^2(n) = E(2\nu^{(1)}(\omega) - n)^2 = E\big[4(\nu^{(1)}(\omega))^2 - 4n\nu^{(1)}(\omega) + n^2\big]$$

$$= 4E(\nu^{(1)}(\omega))^2 - 4nE\nu^{(1)}(\omega) + n^2$$

$$= 4\mathrm{Var}(\nu^{(1)}(\omega)) + 4(E\nu^{(1)}(\omega))^2 - 4nE\nu^{(1)}(\omega) + n^2$$

$$= 4.n.\frac{1}{2}.\frac{1}{2} + 4\left(\frac{n}{2}\right)^2 - 4.n.\left(\frac{n}{2}\right) + n^2 = n.$$

The last relation shows that typical values of $y(n)$ lie in a domain of order $O(\sqrt{n})$. In the next lecture we will give this statement an accurate form.

Lecture 3. De Moivre-Laplace and Poisson Limit Theorems

3.1 De Moivre-Laplace Theorems

3.1.1 Local and Integral De Moivre-Laplace Theorems

We again consider a binomial distribution with probabilities p and q, - i.e. $p_k = \binom{n}{k} p^k q^{n-k}$ and, to fix ideas, we assume that $p > q$. The number k takes values from 0 to n, so we have $n + 1$ probabilities p_k. We now study the question of the behavior of these probabilities as a function of k, for large n. We will see that there is a relatively small domain of values of k (of size $\sqrt{n}$) where the p_k are comparatively large and a remaining domain where the p_k are negligible. In order to define those k for which p_k is large we find k_0 such that $p_{k_0} = \max_k p_k$. We have the relation:

$$\frac{p_{k+1}}{p_k} = \frac{\binom{n}{k+1} p^{k+1} q^{n-k-1}}{\binom{n}{k} p^k q^{n-k}} = \frac{n! \, k! \, (n-k)!}{(k+1)! \, (n-k-1)! \, n!} \frac{p}{q} = \frac{(n-k) \, p}{(k+1) \, q}.$$

We now find for which k the inequality $p_{k+1}/p_k \geq 1$ holds. We have

$$\frac{n-k}{k+1} \frac{p}{q} \geq 1, \qquad (n-k)p \geq q(k+1), \qquad np - q \geq k.$$

Conversely, for $k > np - q$ we have $p_{k+1}/p_k < 1$. So $k_0 = [np - q]$. It is therefore natural to expect that the larger values of the probability p_k occur around the point $k_0 = np$.

We now formulate the De Moivre-Laplace Limit Theorem which reinforces the previous statement. Let A, B be two arbitrary numbers with $A < B$.

Theorem 3.1 (Local Limit Theorem: De Moivre-Laplace). *Let* $np + A\sqrt{n} \leq k \leq np + B\sqrt{n}$. *Then*

$$p_k = \frac{1}{\sqrt{2\pi npq}} e^{\frac{-(k-np)^2}{2npq}} (1 + r_n(k)),$$

where the remainder $r_n(k)$ *converges to* 0 *as* $n \to \infty$ *uniformly in* k *- i.e.*

$$\max_{np + A\sqrt{n} \leq k \leq np + B\sqrt{n}} |r_n(k)| \to 0 \qquad \text{as} \quad n \to \infty.$$

Proof. We use Stirling's formula

$$r! \sim \sqrt{2\pi r}\, r^r e^{-r} \text{ as } r \to \infty.$$

In addition, we use the following simple statements: under the conditions of the theorem

$$\frac{k}{n} \to p, \quad \frac{n-k}{n} \to q \quad \text{as} \quad n \to \infty.$$

We now have

$$
\begin{aligned}
p_k &= \binom{n}{k} p^k q^{n-k} = \frac{n!}{k!\,(n-k)!} p^k q^{n-k} \\
&\sim \frac{\sqrt{2\pi n}\, n^n e^{-n} p^k q^{n-k}}{\sqrt{2\pi k}\, k^k e^{-k} \sqrt{2\pi(n-k)}\,(n-k)^{n-k} e^{-n+k}} \\
&= \frac{1}{\sqrt{2\pi n \frac{k}{n}\left(\frac{n-k}{n}\right)}} \left(\frac{k}{n}\right)^{-k} \left(\frac{n-k}{n}\right)^{-n+k} p^k q^{n-k} \\
&\sim \frac{1}{\sqrt{2\pi npq}} e^{-k \ln \frac{k}{n} - (n-k)\ln \frac{n-k}{n} + k \ln p + (n-k)\ln q}.
\end{aligned}
$$
(3.1)

We set $z = (k - np)/\sqrt{npq}$ and we consider the exponents separately:

$$
\begin{aligned}
S &= -k \ln \frac{k}{n} - (n-k)\ln \frac{n-k}{n} + k \ln p + (n-k)\ln q \\
&= -k \ln\left(\frac{np + z\sqrt{npq}}{n}\right) - (n-k)\ln\left(\frac{nq - z\sqrt{npq}}{n}\right) + k \ln p + (n-k)\ln q \\
&= -k \ln(p + z\sqrt{pq/n}) - (n-k)\ln(q - z\sqrt{pq/n}) + k \ln p + (n-k)\ln q.
\end{aligned}
$$

By Taylor's formula with a Lagrange remainder we have

$$\ln(p + z\sqrt{pq/n}) = \ln p + \frac{z\sqrt{q}}{\sqrt{pn}} - \frac{z^2 q}{2pn} + K_1(p)\frac{z^3 (pq)^{3/2}}{n^{3/2}},$$

$$\ln(q - z\sqrt{pq/n}) = \ln q - \frac{z\sqrt{p}}{\sqrt{qn}} - \frac{z^2 p}{2qn} + K_2(p)\frac{z^3 (pq)^{3/2}}{n^{3/2}},$$

where K_1 and K_2 are the coefficients of the remainders which arise from the third derivatives at intermediate points. We only need the fact that as $n \to \infty$ these constants are bounded by $|K_1|, |K_2| \le C$, where C is a constant that does not depend on n.

$$
\begin{aligned}
S &= -k \ln p - \frac{zk\sqrt{q}}{\sqrt{pn}} + \frac{z^2 qk}{2pn} - (n-k)\ln q + \frac{z\sqrt{p}(n-k)}{\sqrt{qn}} + \frac{z^2 p(n-k)}{2qn} \\
&\quad - K_1(p)\frac{z^3 (pq)^{3/2}}{n^{3/2}} k - K_2(p)\frac{z^3 (pq)^{3/2}}{n^{3/2}}(n-k) - k \ln p - (n-k)\ln q.
\end{aligned}
$$

The expressions which contain $K_1(p)$ and $K_2(p)$ converge uniformly to 0 since z varies between bounded limits, $A \le z \le B$. We now replace k and $(n-k)$ by their expressions in terms of z. Since $k/n \to p$ and $(n-k)/n \to q$ we have

$$S = -\frac{kz\sqrt{q}}{\sqrt{pn}} + \frac{z\sqrt{p}(n-k)}{\sqrt{qn}} + \frac{z^2 q}{2p}\frac{k}{n} + \frac{z^2 p}{2q}\left(\frac{n-k}{n}\right) + o(1)$$

$$= \frac{-(np + z\sqrt{npq})}{\sqrt{pn}}z\sqrt{q} + \frac{(nq - z\sqrt{npq})}{\sqrt{qn}}z\sqrt{q} + \frac{z^2 q}{2} + \frac{z^2 p}{2} + o(1)$$

$$= -\sqrt{npq}\,z - z^2 q + \sqrt{npq}\,z - z^2 p + z^2/2 + o(1) = -z^2/2 + o(1).$$

$\square$

In probability theory statements involving the asymptotic behavior of individual probabilities are sometimes called local limit theorems.

Corollary 3.1. $\max_k p_k \to 0$ *as* $n \to \infty$.

Proof. We saw that $\max p_k$ is obtained for $k_0 = [np - q]$. For such a k_0

$$p_{k_0} \sim \frac{1}{\sqrt{2\pi npq}} \to 0 \text{ as } n \to \infty,$$

since $z_0 = (k_0 - np)/\sqrt{npq}$ and $|z_0| \le 2/\sqrt{npq} \to 0$ as $n \to \infty$. $\square$

Theorem 3.2 (Integral Limit Theorem: De Moivre-Laplace). *Let* A, B *be fixed numbers with* $A < B$. *Then*

$$\lim_{n \to \infty} \sum_{\substack{np + A\sqrt{npq} \le k \\ \le np + B\sqrt{npq}}} p_k = \frac{1}{\sqrt{2\pi}} \int_A^B e^{-z^2/2}\, dz.$$

Proof. Again we set $z = (k - np)/\sqrt{npq}$. When k takes all integer values such that $0 \le k \le n$ in turn, z varies between the limits $-\sqrt{np/q} \le z \le \sqrt{np/q}$ by steps of length $1/\sqrt{npq}$. By Theorem 3.1 we have

$$\sum_{\substack{np + A\sqrt{npq} \le k \\ np + B\sqrt{npq}}} p_k = \sum_{A \le z \le B} \frac{1}{\sqrt{2\pi npq}} e^{-z^2/2} (1 + o(1)).$$

The sum

$$\sum_{A \le z \le B} \frac{1}{\sqrt{2\pi npq}} e^{-z^2/2}$$

is a Riemann sum for the integral $(1/\sqrt{2\pi}) \int_A^B e^{-z^2/2}\, dz$, and the remainder $o(1)$ converges to 0 uniformly with respect to all considered values of k. $\square$

The expression on the right hand side of the statement of the theorem is a probability calculated with the Gaussian density $p(z) = (1/\sqrt{2\pi})e^{-z^2/2}$. This density plays an important role in many problems of probability theory.

Corollary 3.2. *For any A*

$$\lim_{n \to \infty} \sum_{k \le np + A \sqrt{npq}} p_k = \frac{1}{\sqrt{2\pi}} \int_{-\infty}^{A} e^{-z^2/2} \, dz,$$

$$\lim_{n \to \infty} \sum_{k \ge np + A \sqrt{npq}} p_k = \frac{1}{\sqrt{2\pi}} \int_{A}^{+\infty} e^{-z^2/2} \, dz.$$

We show only the first statement, as the second statement is proved similarly. Let $\epsilon > 0$ be arbitrary. Since $\frac{1}{\sqrt{2\pi}} \int_{-\infty}^{\infty} e^{-z^2/2} \, dz = 1$, there exists a $B > 0$ such that $\frac{1}{\sqrt{2\pi}} \int_{-B}^{B} e^{-z^2/2} \, dz = 1 - \epsilon/4$. By Theorem 3.1 we have

$$\lim_{n \to \infty} \sum_{\substack{np - B\sqrt{npq} \le k \\ \le np + B\sqrt{npq}}} p_k = \frac{1}{\sqrt{2\pi}} \int_{-B}^{B} e^{-z^2/2} \, dz = 1 - \epsilon/4.$$

Therefore for all $n \ge n_0(\epsilon)$

$$\sum_{np - B\sqrt{npq} \le k \le np + B\sqrt{npq}} p_k \ge 1 - \epsilon/2.$$

Consequently, for such n

$$\sum_{k < np - B\sqrt{npq}} p_k \le \epsilon/2.$$

We therefore have

$$\Sigma = \sum_{k \le np + A\sqrt{npq}} p_k = \sum_{k < np - B\sqrt{npq}} p_k + \sum_{\substack{np - B\sqrt{npq} \le k \\ \le np + A\sqrt{npq}}} p_k = \Sigma_1 + \Sigma_2.$$

We already showed that $\Sigma_1 \le \epsilon/2$ for $n \ge n_0(\epsilon)$. By the theorem we have proved, Σ_2 satisfies the following equalities:

$$\lim_{n \to \infty} \Sigma_2 = \frac{1}{\sqrt{2\pi}} \int_{-B}^{A} e^{-z^2/2} \, dz = \frac{1}{\sqrt{2\pi}} \int_{-\infty}^{A} e^{-z^2/2} \, dz - \epsilon/8.$$

Therefore, for $n > n_1(\epsilon)$

$$\left| \Sigma_2 - \frac{1}{\sqrt{2\pi}} \int_{-\infty}^{A} e^{-z^2/2} \, dz + \epsilon/8 \right| \le \epsilon/8.$$

So for $n \ge \max(n_0(\epsilon), n_1(\epsilon))$

$$\left| \Sigma - \frac{1}{\sqrt{2\pi}} \int_{-\infty}^{A} e^{-z^2/2} \, dz \right| \le \left| \Sigma_2 - \frac{1}{\sqrt{2\pi}} \int_{-\infty}^{A} e^{-z^2/2} \, dz + \epsilon/8 \right| + \epsilon/8 + |\Sigma_1|$$

$$\le \epsilon/8 + \epsilon/8 + \epsilon/2 < \epsilon.$$

3.1.2 Application to Symmetric Random Walks

Let $y(n)$ be the coordinate of a random walk point at time n. We have already seen that $y(n) = 2\nu^{(1)}(\omega) - n$. Recall that $\nu^{(1)}(\omega)$ has a binomial distribution with $p = q = 1/2$. Therefore

$$
\begin{aligned}
P(A\sqrt{n} \le y(n) \le B\sqrt{n}) &= P(A\sqrt{n} \le 2\nu^{(1)}(\omega) - n \le B\sqrt{n}) \\
&= P(A\sqrt{n}/2 + n/2 \le \nu^{(1)}(\omega) \le B\sqrt{n}/2 + n/2) \\
&= P\left(A\sqrt{n \cdot \frac{1}{2} \cdot \frac{1}{2}} + n/2 \le \nu^{(1)}(\omega) \le B\sqrt{n \cdot \frac{1}{2} \cdot \frac{1}{2}} + n/2\right) \\
&\to \frac{1}{\sqrt{2\pi}} \int_A^B e^{-z^2/2}\, dz.
\end{aligned}
$$

The last relation shows that for any interval $[A, B]$ the probability that $y(n)/\sqrt{n} \in [A, B]$ converges to a positive limit as $n \to \infty$. This means that for large n, in typical cases (i.e. with positive probability) the random walk takes values of order $\sqrt{n}$, i.e. the random walk point at time n has departed from its initial value by a distance of order $\sqrt{n}$. Such a feature is typical of diffusion processes.

3.1.3 A Problem in Mathematical Statistics

We now consider a problem in mathematical statistics which is related to a sequence of n identical trials with two outcomes. For convenience we call these outcomes "success" and "failure" and denote them respectively by 1 and 0. The result of n trials therefore has the form $\omega = \{x_1, \ldots, x_n\}$ where each x_i takes the values 1 or 0 with probabilities p and $q = 1 - p$ respectively, and $\sum_{i=1}^n x_i$ is the number of successes. A fundamental problem of mathematical statistics in this case consists in estimating the value of p from a given concrete ω. For example, if we are dealing with a real-life game situation, and if the game is fair, i.e. theoretically $p = q = 1/2$, then an estimator of the difference $p - 1/2$ is an estimator of whether cheating has occurred.

On the basis of the law of large numbers, a natural candidate for the estimator of p is $\nu^{(1)}(\omega)/n = (1/n) \sum_{i=1}^n x_i$, i.e. the proportion of successes. In connection with this we run into a general problem of mathematical statistics, that of the accuracy of a statistical estimator. The usual approach applied to our problem consists in the following. We start from the fact that a probability distribution is given on the space Ω for a sequence of independent trials with an unknown probability p. We choose $\alpha > 0$ and call it the tolerance level, and we find a number $a(p, n)$ such that

$$
P\big\{(\omega|\,|\frac{\nu^{(1)}(\omega)}{n} - p| \le a(p, n))\big\} \ge 1 - \alpha. \tag{3.2}
$$

The last inequality means that

$$
-a(p, n)n \le \nu^{(1)}(\omega) - pn \le a(p, n)n,
$$

or

$$pn - a(p,n)n \leq \nu^{(1)}(\omega) \leq pn + a(p,n)n. \qquad (3.3)$$

We now consider the following functions of p, which depend on n:

$$f_+(p) = pn + a(p,n)n,$$

$$f_-(p) = pn - a(p,n)n.$$

We assume that $f_\pm(p)$ are strictly monotonous in p and that there exist inverse functions $g_+(p)$ and $g_-(p)$, i.e. $g_+(f_+(p)) = p$ and $g_-(f_-(p)) = p$. It then follows that equation (3.3) can be rewritten in the following way:

$$g_+\left(\nu^{(1)}(\omega)\right) \leq p \leq g_-\left(\nu^{(1)}(\omega)\right). \qquad (3.4)$$

The interpretation of (3.4) is that with probability at least $1-\alpha$, p lies between the limits

$$g_+\left(\nu^{(1)}(\omega)\right) \leq p \leq g_-\left(\nu^{(1)}(\omega)\right). \qquad (3.5)$$

Resorting to the frequency interpretation of probability we can say that an estimation of p by way of (3.5) is correct a proportion $(1-\alpha)$ of the time. In other words if we carry out an estimation of p by using a series of n trials N times, then in approximately $N(1-\alpha)$ cases the estimation (3.4) is correct, and in the remaining cases it is incorrect. As we will see, the smaller α is, the worse the accuracy of the estimator, i.e. the greater is $g_-\left(\nu^{(1)}(\omega)\right) - g_+\left(\nu^{(1)}(\omega)\right)$. These circumstances dictate the choice of the number α. If the frequency of errors must be minimal, because each error can lead to serious consequences (as for example in the case of medical diagnosis), then α will be chosen very small, but then the accuracy of the estimator is worse. If an estimation error is permissible in a given proportion of cases, then we can choose a greater α, and obtain a tighter interval for p. The prevailing values of α are $\alpha = 0.1, 0.05, 0.01, 0.005$ and 0.001.

We now find an explicit form for f_+, f_-, g_+ and g_-. An exact solution of (3.2) is difficult to obtain. Let us assume that n is large enough for us to be able to use the De Moivre-Laplace Limit Theorem. Then

$$P\left\{(\omega|\,\left|\frac{\nu^{(1)}(\omega)}{n} - p\right| \leq \frac{a\sqrt{pq}}{\sqrt{n}})\right\}$$

$$= P\left\{(\omega|\; -a\sqrt{np(1-p)} \leq \nu^{(1)}(\omega) - np \leq a\sqrt{np(1-p)})\right\}$$

$$= P\left\{\omega|\; pn - a\sqrt{np(1-p)} \leq \nu^{(1)}(\omega) \leq pn + a\sqrt{np(1-p)}\right\}$$

$$\approx \frac{1}{\sqrt{2\pi}} \int_{-a}^{a} e^{-z^2/2}\, dz = 1 - \alpha.$$

Given the number α we can find $a = a_\alpha$ by using tables for the normal distribution. Then

$$f_+(p) = pn + a_\alpha\sqrt{np(1-p)}, \quad f_-(p) = pn - a_\alpha\sqrt{np(1-p)}.$$

Let us set $f_+(p) = t$. We have $t - pn = a_\alpha\sqrt{np(1-p)}$,

$$t^2 - 2pnt + p^2 n^2 - a_\alpha^2 np(1-p) = 0.$$

Analogous calculations hold for $f_-(p)$. If we solve these equations for p we obtain p as a function of t. The two different choices of roots correspond to the two functions $g_\pm$. In the plane (p,t) the equation $t^2 - 2pnt + p^2 n^2 - a_\alpha^2 np(1-p) = 0$ is the equation of an ellipse passing through the points $(0,0)$ and $(1,n)$. It is convenient to normalize t – that is to set $\tau = t/n$.

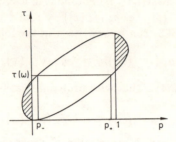

Fig. 3.1.

We then obtain $\tau^2 - 2p\tau + p^2 - a_\alpha^2 p(1-p)/n = 0$. It is now clear that the width of this ellipse is of the order $O(a_\alpha/\sqrt{n})$. The smaller α is, the greater a_α is and the wider the ellipse is. The ellipse must be used in the following way: given ω we find the quantity $\tau = \tau(\omega) = \nu^{(1)}(\omega)/n$ and draw in the plane (p,τ) a horizontal line at height $\tau(\omega)$. The line intersects the ellipse at the two points p_-, p_+. Then $p_- \leq p \leq p_+$ with probability at least $1 - \alpha$. The fact that the ellipse intersects the "forbidden domain" $p < 0$ and $p > 1$ is connected with the fact that the approximation by a Gaussian distribution turns out to be invalid here.

3.1.4 Generalizations of the De Moivre-Laplace Theorem

We now consider two generalizations of Theorem 3.1, which will be of use to us later in our study of higher-dimensional random walks. Let $X = \{a_1, a_2; b_1, b_2\}$, $p(a_1) = p(a_2) = p$, $p(b_1) = p(b_2) = q$, $2p + 2q = 1$ and $X = \{a_1, a_2; b_1, b_2; c_1, c_2\}$, $p(a_1) = p(a_2) = p$, $p(b_1) = p(b_2) = q$, $p(c_1) = p(c_2) = r$, $2p + 2q + 2r = 1$. We consider the events $C_n = \{\omega | \nu^{(1)}(\omega) = \nu^{(2)}(\omega), \nu^{(3)}(\omega) = \nu^{(4)}(\omega)\}$ in the first case; $C_n = \{\omega | \nu^{(1)}(\omega) = \nu^{(2)}(\omega), \nu^{(3)}(\omega) = \nu^{(4)}(\omega), \nu^{(5)}(\omega) = \nu^{(6)}(\omega)\}$ in the second case.

Theorem 3.3. *The following relations hold:*

$$\lim_{n \to \infty} nP(C_n) = \frac{1}{2\pi\sqrt{pq}} \quad \text{in the first case, and}$$

$$P(C_n) \leq \frac{\text{const}}{n^{3/2}} \quad \text{in the second case,}$$

where const *is a constant which depends on* p, q *and* r.

Proof. Since $\sum_i \nu^{(2i)}(\omega) = \sum_i \nu^{(2i+1)}(\omega)$ for $\omega \in C_n$, we can assume that n is even. In the first case

$$C_n = \bigcup_k C_{n,k}$$

with

$$C_{n,k} = \{\omega | \nu^{(1)}(\omega) = \nu^{(2)}(\omega) = k, \nu^{(3)}(\omega) = \nu^{(4)}(\omega) = n/2 - k\},$$

and in the second case

$$C_n = \bigcup_{k_1,k_2} C_{n,k_1,k_2},$$

with

$$C_{n,k_1,k_2} = \{\omega | \nu^{(1)}(\omega) = \nu^{(2)}(\omega) = k_1, \nu^{(3)}(\omega) = \nu^{(4)}(\omega) = k_2,$$
$$\nu^{(5)}(\omega) = \nu^{(6)}(\omega) = n/2 - k_1 - k_2\}.$$

For each $\omega \in C_{n,k}$ clearly $p(\omega) = p^{2k} . q^{n-2k}$. Therefore $P(C_{n,k}) = |C_{n,k}| . p^{2k} . q^{n-2k}$. In order to calculate $|C_{n,k}|$ we note that $|C_{n,k}|$ is equal to the number of ways in which we can choose from the set $\{1, 2, \ldots, n\}$ two subsets with k elements and two subsets with $n/2 - k$ elements. The first subset can be chosen in $\binom{n}{k}$ ways. When the first subset is chosen then for the choice of the second subset we have $\binom{n-k}{k}$ possibilities. Once we have chosen the first two subsets, there are $\binom{n-2k}{n/2-k}$ possibilities for the choice of the third one. Finally

$$|C_{n,k}| = \binom{n}{k}\binom{n-k}{k}\binom{n-2k}{n/2-k}$$
$$= \frac{n!(n-k)!(n-2k)!}{k!(n-k)!k!(n-2k)!((n/2-k)!)^2} = \frac{n!}{(k!)^2((n/2-k)!)^2}.$$

By analogous arguments we get, in the second case,

$$|C_{n,k_1,k_2}| = \binom{n}{k_1}\binom{n-k_1}{k_1}\binom{n-2k_1-k_2}{k_2}\binom{n-2k_1-2k_2}{n/2-k_1-k_2}$$
$$= \frac{n!}{(k_1!)^2(k_2!)^2((n/2-k_1-k_2)!)^2}.$$

We now have

$$P(C_{n,k}) = \frac{n!}{(k!)^2((n/2-k)!)^2}p^{2k}q^{n-2k}$$
$$= \frac{n!}{(2k)!(n-2k)!}(2p)^{2k}(2q)^{n-2k}\frac{1}{2^n}\frac{(2k)!}{(k!)^2}\frac{(n-2k)!}{((n/2-k)!)^2}.$$

We note that $\frac{1}{2^{2k}}\frac{(2k)!}{(k!)^2}$ is the probability of obtaining k ones in a sequence of $2k$ trials with $p = q = 1/2$. By Theorem 3.1

$$\frac{1}{2^{2k}}\frac{(2k)!}{(k!)^2} \sim \frac{1}{\sqrt{2\pi 2k\frac{1}{2}\frac{1}{2}}} = \frac{1}{\sqrt{\pi k}}, \ k \to +\infty.$$

We also have

$$\frac{(n-2k)!}{\left((n/2-k)!\right)^2 2^{n-2k}} \sim \frac{1}{\sqrt{\pi(n/2-k)}}, \ n/2-k \to +\infty.$$

In all cases

$$\frac{1}{2^{2k}}\frac{(2k)!}{(k!)^2}\frac{1}{2^{n-2k}}\frac{(n-2k)!}{\left((n/2-k)!\right)^2} \le \frac{c}{\sqrt{n}}$$

for some absolute constant c. So for arbitrary $\delta > 0$ we now have

$$P(C_n) = \sum_k P(C_{n,k}) = \sum_k p_{2k}\left(\frac{1}{2^{2k}}\frac{(2k)!}{(k!)^2}\right)\left(\frac{1}{2^{n-2k}}\frac{(n-2k)!}{((n/2-k)!)^2}\right)$$

$$= \Big(\sum_{|k/n-p|<\delta} + \sum_{|k/n-p|\ge\delta}\Big)p_{2k}\left(\frac{1}{2^{2k}}\frac{(2k)!}{(k!)^2}\right)\left(\frac{1}{2^{n-2k}}\frac{(n-2k)!}{((n/2-k)!)^2}\right)$$

$$= \Sigma_1 + \Sigma_2.$$

Here the p_i denote the probabilities of a binomial distribution with probabilities $2p$ and $2q$. By Chebyshev's inequality $\sum_2$ satisfies

$$\Sigma_2 \le \frac{c}{\sqrt{n}}\sum_{|k/n-p|\ge\delta}p_{2k} \le \frac{c}{\sqrt{n}}\sum_{|2k/n-2p|\ge 2\delta}p_{2k}$$

$$\le \frac{c}{\sqrt{n}}\frac{4nqp}{4\delta^2 n^2} = \frac{cq}{n^{3/2}}, \qquad \Sigma_2 \cdot n \to 0 \quad \text{as} \quad n \to \infty.$$

We can write $\sum_1$ in the following way:

$$\Sigma_1 = \sum_{|k/n-p|<\delta}p_{2k}\left(\frac{1}{2^{2k}}\frac{(2k)!}{(k!)^2}\right)\left(\frac{1}{2^{n-2k}}\frac{(n-2k)!}{\left((n/2-k)!\right)^2}\right)$$

$$\le (1+\alpha)\frac{1}{\sqrt{\pi np}}\frac{1}{\sqrt{\pi nq}}\sum_k p_{2k},$$

where $\alpha = \alpha(\delta) \to 0$ as $\delta \to 0$. The sum $\sum_i p_i = 1$ and the sum $\sum_k p_{2k} \to 1/2$, by Theorem 3.1 (we take only half the terms in the Riemann sum; the exact proof is left to the reader as an exercise). Therefore

$$n\Sigma_1 \le (1+\alpha)\frac{1}{2\pi\sqrt{pq}}.$$

Since α was arbitrary, we have proved that

$$\varlimsup_{n\to\infty} n\Sigma_1 \le \frac{1}{2\pi\sqrt{pq}}.$$

A lower bound is obtained similarly: for $|k/n-p|<\delta$

$$\frac{1}{2^{2k}}\frac{(2k)!}{(k!)^2}\frac{1}{2^{n-2k}}\frac{(n-2k)!}{((n/2-k)!)^2} \geq (1-\alpha)\frac{1}{\sqrt{\pi np}}\frac{1}{\sqrt{\pi nq}},$$

and the fact that $\sum_{|k/n-p|<\delta} p_{2k} \to 1/2$ as $n \to \infty$ can again be proved by using Theorem 3.1.

We now turn to the second case. We have

$$P(C_n) = \sum_{k_1,k_2} P(C_{n,k_1,k_2})$$

$$= \sum_{k_1,k_2} \frac{n! p^{2k_1} q^{2k_2} r^{n-2k_1-2k_2}}{(k_1!)^2 (k_2!)^2 ((n/2-k_1-k_2)!)^2}$$

$$= \sum_{k_1} \frac{n! p^{2k_1} (2(q+r))^{n-2k_2}}{(k_1)!(n-2k_1)!}.$$

$$\sum_{k_2} \frac{(n-2k_1)!}{(k_1!)^2 ((n/2-k_1-k_2)!)^2} \left(\frac{q}{2(q+r)}\right)^{2k_2} \left(\frac{r}{2(q+r)}\right)^{n-2k_1-2k_2}.$$

Here $\sum_{k_2}$ is equal to the sum which we estimated in the first case, therefore for a certain constant c_1

$$\sum_{k_2} \frac{(n-2k_1)!}{(k_1!)^2 ((n/2-k_1-k_2)!)^2} \left(\frac{q}{2(q+r)}\right)^{2k_2} \left(\frac{r}{2(q+r)}\right)^{n-2k_1-2k_2}$$

$$\leq \frac{c_1}{n-2k_1+1}.$$

It follows that

$$P(C_n) \leq c_1 \sum_{k_1} \frac{n!(2p)^{2k_1} (2(q+r))^{n-2k_2}}{(k_1!)^2 (n-2k_1)!(n-2k_1+1)2^{2k_1}}$$

$$= c_1 \sum_{k_1} \frac{n!(2p)^{2k_1} (2(q+r))^{n-2k_2}}{(2k_1)!(n-2k_1)!} \frac{(2k_1)!}{((k_1)!)^2 2^{2k_1} (n-2k_1+1)}.$$

Moreover, as previously we can write

$$\frac{(2k_1)!}{(k_1!)^2 2^{2k_1}} \leq \frac{c_1}{\sqrt{k_1+1}}.$$

If we denote by p'_i the probabilities of a binomial distribution with probabilities $2p$ and $2(q+r)$ we obtain

$$P(C_n) \leq c_1^2 \sum_{k_1} p'_{2k_1} \frac{1}{\sqrt{k_1+1}(n-2k_1+1)}.$$

Also, as in the first case, one can show that the main contribution to this sum is given by those terms where $k_1/n \sim p$. For such k_1 we have

$$\frac{1}{\sqrt{k_1+1}}\frac{1}{(n-2k_1+1)} = \frac{1}{n^{3/2}}\frac{1}{\sqrt{k_1/n+1/n}}\frac{1}{(1-2k_1/n+1/n)}$$

$$\sim \frac{1}{n^{3/2}}\frac{1}{\sqrt{2p2(q+r)}}.$$

The desired statement then follows easily. $\square$

Later we consider other methods which lead to similar estimates.

3.2 The Poisson Distribution and the Poisson Limit Theorem

3.2.1 Poisson Limit Theorem

The Poisson probability distribution is the following:

$$p_k = \begin{cases} 0, & k < 0; \\ e^{-\lambda}\frac{\lambda^k}{k!}, & k \geq 0. \end{cases}$$

Here $\lambda > 0$ is the parameter of the Poisson distribution and k takes its values in the set of integers $-\infty < k < \infty$. If a random variable ξ has a Poisson distribution with parameter λ then

$$E\xi = \sum_{k\geq 0} kp_k = \sum_{k\geq 0} ke^{-\lambda}\frac{\lambda^k}{k!} = e^{-\lambda}\sum_{k\geq 1}\frac{\lambda^k}{(k-1)!}$$

$$= e^{-\lambda}\lambda\sum_{k\geq 1}\frac{\lambda^{k-1}}{(k-1)!} = e^{-\lambda}\lambda\sum_{t\geq 0}\frac{\lambda^t}{t!} = \lambda.$$

We now consider a problem in which a Poisson distribution arises. Let $X = \{1,0\}$ and $\omega = \{x_1,\ldots,x_n\}$, $x_i \in X$. We assume, given on Ω, a probability distribution which corresponds to a sequence of n identical independent trials but where $p = p(1)$ depends on n – i.e. $p = p_n$ – in such a way that as $n \to \infty$ $\lim_{n\to\infty} np_n = \lambda$. Then $E\nu^{(1)} = np_n \to \lambda$ as $n \to \infty$.

Theorem 3.4 (Poisson Limit).

$$\lim_{n\to\infty} P\{\nu^{(1)} = k\} = e^{-\lambda}\frac{\lambda^k}{k!}.$$

Proof. We have

$$P(\nu^{(1)} = k) = \binom{n}{k}p_n^k(1-p_n)^{n-k}$$

$$= \frac{n(n-1)\ldots(n-k+1)}{k!}p_n^k e^{(n-k)\ln(1-p_n)}.$$

In our situation k is fixed but $n \to \infty$. Therefore $(n - k) \ln(1 - p_n) \sim -(n - k)p_n = -p_n n(1 - k/n) \to -\lambda$ as $n \to \infty$. Furthermore

$$n(n - 1) \ldots (n - k + 1)p_n^k = (np_n)^k (1 - 1/n)(1 - 2/n) \ldots$$
$$\ldots (1 - (k - 1)/n) \to \lambda^k \qquad \text{as} \quad n \to \infty.$$

Finally $P(\nu^{(1)} = k) \to \frac{\lambda^k}{k!}e^{-\lambda}$ as $\quad \to \infty$. $\qquad\qquad \square$

The condition $p_n \sim \lambda/n$ implies that the probability of 1 (of success) is very small and decreases as $O(1/n)$. It also means that $E\nu^{(1)} \sim \lambda$ as $n \to \infty$ and remains finite – i.e. the total number of ones in a word of length n on average remains finite and does not increase with n. For that reason the Poisson Limit Theorem is sometimes called the limit theorem for rare events.

3.2.2 Application to Statistical Mechanics

Here we describe an application of the Poisson Limit Theorem to statistical mechanics. An ideal gas in a volume V is a system of N non-interacting particles. We assume that V is a d-dimensional cube centered at 0 and with sides R and that $R \to \infty$ and $N \to \infty$ in such as way that the number of particles in a unit volume remains finite, $N/R^d \to \lambda > 0$ as $R \to \infty$, $N \to \infty$. Such a limit transition is sometimes called thermodynamic. The absence of interaction appears in the assumption that each particle of gas is uniformly distributed in V independently of the remaining particles. More precisely this means that $X = V$ and $\Omega = X^{[1,N]}$. Also a uniform distribution is given on V for which $p(Q) = \mathrm{vol}\, Q/R^d$, where $\mathrm{vol}\, Q$ is the volume of $Q \subset V$, and a probability distribution is given on Ω for a sequence of N trials. The domain Q is now fixed, and we introduce the random variable $\nu^{(Q)}(\omega)$ equal to the number of particles which lie in Q.

Theorem 3.5.

$$\lim P(\nu^{(Q)}(\omega) = k) = \frac{(\lambda \mathrm{vol}\, Q)^k}{k!}e^{-\lambda \mathrm{vol}\, Q}.$$

Proof. Let us fix indices $i_1, \ldots, i_k$ for the particles which lie in Q, and let $C_{i_1, \ldots, i_k} = \{\omega | x_{i_s} \in Q,\ 1 \le s \le k,\ x_j \notin Q \text{ for } j \ne i_1, \ldots, i_k\}$. Then

$$P(\nu^{(Q)}(\omega) = k) = \sum P(C_{i_1, \ldots, i_k}).$$

Each $C_{i_1, \ldots, i_k}$ is a cylinder and for it

$$P(C_{i_1, \ldots, i_k}) = \left(\frac{\mathrm{vol}\, Q}{R^d}\right)^k \left(1 - \frac{\mathrm{vol}\, Q}{R^d}\right)^{N-k}.$$

Therefore

$$P(\nu^{(Q)}(\omega) = k) = \binom{n}{k}\left(\frac{\text{vol}\,Q}{R^d}\right)^k\left(1 - \frac{\text{vol}\,Q}{R^d}\right)^{N-k}.$$

In other words we have a binomial distribution in which the probability of 1 (success consists in the particle lying in Q) is equal to $p_n = \text{vol}\,Q/R^d$. Therefore $P_N N = N\text{vol}\,Q/R^d \to \lambda\text{vol}\,Q$, and the desired result follows from Theorem 3.4.   □

Theorem 3.5 shows that the number of particles of an ideal gas lying in a given fixed domain Q, when passing to the limit as previously described, follows a Poisson distribution with parameter $\lambda\text{vol}\,Q$.

Lecture 4. Conditional Probability and Independence

4.1 Conditional Probability and Independence of Events

Let $(\Omega, \mathcal{F}, P)$ be a probability space and $A \in \mathcal{F}$, $B \in \mathcal{F}$ be two events, with $P(B) > 0$.

Definition 4.1. The conditional probability of an event A given an event B is given by

$$P(A|B) = \frac{P(A \cap B)}{P(B)}.$$

The importance of the concept of conditional probability lies in the fact that for a large number of problems the initial data consist of conditional probabilities from which one wishes to find properties of ordinary non-conditional probabilities. In this and the next three lectures we consider examples of such problems.

It is clear that the conditional probability $P(A|B)$ depends on A and on B, but this dependence has a very different nature. As a function of A the conditional probability satisfies the usual properties of probability:

1. $P(A|B) \geq 0$;
2. $P(\Omega|B) = 1$;
3. for a sequence of disjoint events $\{A_i\}$ (i.e. $A_i \cap A_j = \emptyset$ for $i \neq j$), with $A = \bigcup_i A_i$

$$P(A|B) = \sum_i P(A_i|B).$$

The nature of the dependence on B arises from the formula of total probability which we give below. Let $\{B_1, B_2, \ldots, B_n, \ldots\}$ be a finite or countable partition of the space Ω – i.e. a collection of sets $\{B_i\}$ such that $B_i \cap B_j = \emptyset$ and $\bigcup_i B_i = \Omega$. We also assume that $P(B_i) > 0$ for every i. For any $A \in \mathcal{F}$ we have

$$P(A) = \sum_i P(A \cap B_i) = \sum_i P(A|B_i)P(B_i).$$

The relation

$$P(A) = \sum P(A|B_i)P(B_i) \tag{4.1}$$

is called the total probability formula. The nature of this formula is reminiscent of expressions of the type of a multiple integral written in terms of iterated integrals. The conditional probability $P(A|B_i)$ plays the role of the inner integral and the summation over i is the analog of the outer integral.

Sometimes in mathematical statistics the events B_i are called hypotheses, since the choice of B_i defines a probability distribution on $\mathcal{F}$, and probabilities $P(B_i)$ are called prior probabilities (i.e. given before the experiment). We assume that as a result of the trial an event A occurred and we wish, on the basis of this, to draw conclusions on which of the hypotheses B_i is most likely. This estimation is done by the calculation of the probabilities $P(B_k|A)$, which sometimes are called posterior (after the experiment) probabilities. We have

$$P(B_k|A) = \frac{P(B_k \cap A)}{P(A)} = \frac{P(A|B_k).P(B_k)}{\sum_k P(B_k)P(A|B_k)}. \tag{4.2}$$

The relation (4.2) is called Bayes' formula.

We now introduce one of the central concepts of probability theory. Assume that $P(B) > 0$. We say that the event A does not depend on the event B if $P(A|B) = P(A)$. It follows from the expression for conditional probability that $P(A \cap B) = P(A)P(B)$. This latter relation is taken as the definition of independence in the general case. This also shows that if $P(A) > 0$ the independence of B from A follows from the independence of A from B.

Definition 4.2. The events A, B are said to be independent if $P(A \cap B) = P(A).P(B)$.

Lemma 4.1. *The events A and B are independent if and only if the events in any of the pairs $(\bar{A}, B)$, $(A, \bar{B})$ or $(\bar{A}, \bar{B})$ are independent.*

Proof. Let us prove for example the equality $P(\bar{A} \cap \bar{B}) = P(\bar{A}).P(\bar{B})$. We have

$$P(\bar{A} \cap \bar{B}) = P((\Omega \setminus A) \cap \bar{B}) = P(\bar{B}) - P(A \cap \bar{B})$$
$$= P(\bar{B}) - P(A \cap (\Omega \setminus B)) = P(\bar{B}) - P(A) + P(A \cap B)$$
$$= P(\bar{B}) - P(A) + P(A)P(B) = P(\bar{B}) - P(A)(1 - P(B))$$
$$= P(\bar{B}) - P(A)P(\bar{B}) = P(\bar{B})(1 - P(A)) = P(\bar{B})P(\bar{A}).$$

The remaining equalities are proved in an analogous way. □

The statements of Lemma 4.1 can be reformulated differently. Consider two partitions of the space Ω, namely $(A, \bar{A})$ and $(B, \bar{B})$. We saw earlier (Lecture 1) that finite partitions correspond uniquely to finite subalgebras. Let us denote these subalgebras, in our case, by $\mathcal{F}_A$ and $\mathcal{F}_B$. It follows from Lemma 4.1 that $P(C_1 \cap C_2) = P(C_1).P(C_2)$ for any $C_1 \in \mathcal{F}_A$, $C_2 \in \mathcal{F}_B$. We now directly generalize Definition 4.2.

Definition 4.3. Let $\mathcal{F}_1$ and $\mathcal{F}_2$ be two σ-subalgebras of the σ-algebra $\mathcal{F}$. Then $\mathcal{F}_1$ and $\mathcal{F}_2$ are said to be independent if for any $C_1 \in \mathcal{F}_1$, $C_2 \in \mathcal{F}_2$,

$$P(C_1 \cap C_2) = P(C_1).P(C_2).$$

Definition 4.4. Let $\mathcal{F}_1, \mathcal{F}_2, \ldots, \mathcal{F}_n$ be a finite collection of σ-subalgebras of the σ-algebra $\mathcal{F}$. Then the σ-algebras $\mathcal{F}_i$, $1 \leq i \leq n$ are said to be jointly independent if for any $C_1 \in \mathcal{F}_1, C_2 \in \mathcal{F}_2, \ldots, C_n \in \mathcal{F}_n$

$$P(C_1 \cap C_2 \cap \ldots \cap C_n) = P(C_1)P(C_2)\ldots P(C_n).$$

It is clear that one can find σ-algebras which are pairwise independent but not jointly independent.

4.2 Independent σ-algebras and sequences of independent trials

Let Ω consist of the words $\omega = \{x_1, \ldots, x_n\}$, $x_i \in X$ and P be a probability distribution on Ω corresponding to a sequence of independent identical trials. Given $1 \leq i \leq n$ let us introduce the partition ξ_i in which $\omega' = \{x'_1, \ldots, x'_n\}$ and $\omega'' = \{x''_1, \ldots, x''_n\}$ lie in the same element of the partition if and only if $x'_i = x''_i$. By the same token an element C_{ξ_i} of the partition ξ_i is given by a point $x \in X$ and consists of those ω for which $\omega_i = x$. The quotient-space $\Omega|\xi_i$ is canonically isomorphic to X, and the σ-algebra $\mathcal{F}|\xi_i$ is isomorphic to the σ-algebra $\mathcal{B}$ of subsets of the space X (see Lecture 2). The induced probability distribution on $(\Omega|\xi_i, \mathcal{F}(\xi_i))$ agrees with P. The definition of a sequence of independent identical trials means that all σ-algebras $\mathcal{F}(\xi_i)$ are jointly independent and the probability distribution of each of them does not depend on i (identical trials). In the general case of independent σ-subalgebras $\mathcal{F}_i = \mathcal{F}(\xi_i)$ the matter reduces to a sequence of independent identical trials where the space of values ω_i is $(\Omega|\xi_i, \mathcal{F}_i, p_i)$, where p_i is the induced distribution on the measurable space $(\Omega|\xi_i, \mathcal{F}_i)$.

Definition 4.5. Let $\eta_1 = f_1(\omega), \ldots, \eta_n = f_n(\omega)$ be a collection of n random variables. The $\eta_1, \ldots, \eta_n$ are said to be jointly independent random variables if for any Borel subsets $C_1, \ldots, C_n$

$$P\{\eta_1 = f_1(\omega) \in C_1, \ldots, \eta_n = f_n(\omega) \in C_n\} = \prod_{i=1}^{n} P(f_i(\omega) \in C_i).$$

We now show the meaning of this definition when the random variables η_i take a finite or countable number of values. Let $\{a_j^{(i)}\}$ be the values of the random variable η_i and $B_j^{(i)} = \{\omega | \eta_i = a_j^{(i)}\}$. Then for any i the $\{B_j^{(i)}\}$ form a partition of the space Ω which generates a σ-algebra that we will denote by $\mathcal{F}_i$, $1 \leq i \leq n$. It follows from Definition 4.4 that

$$P\{\eta_1 = a_{j_1}^{(1)}, \ldots, \eta_n = a_{j_n}^{(n)}\} = \prod_{i=1}^n P\{\eta_i = a_{j_i}^{(i)}\}.$$

The joint independence of random variables is equivalent to the joint independence of the σ-algebras $\mathcal{F}_i$ which correspond to them.

If the random variables $\eta_1, \ldots, \eta_n$ have probability densities $p_1(x), \ldots$ $\ldots, p_n(x)$ it then follows from Definition 4.4 that

$$P\{a_1 \leq \eta_1 \leq b_1, \ldots, a_n \leq \eta_n \leq b_n\} = \prod_{k=1}^n \int_{a_k}^{b_k} p_k(x)\, dx.$$

Definition 4.5 is of course a special case of Definition 4.4. To see this we take as the σ-algebra $\mathcal{F}_k$ the σ-algebra of the subsets of the form $f_k^{-1}(C)$, where C is a Borel subset.

Theorem 4.1. *Let η_1 and η_2 be independent random variables such that $E\eta_i$ exists, $i = 1, 2$. Then $E(\eta_1 \cdot \eta_2)$ exists and $E(\eta_1 \cdot \eta_2) = E\eta_1 \cdot E\eta_2$.*

We prove this theorem only in the case where η_1 and η_2 take no more than a countable number of values. Let the values of η_1 be the numbers $a_1, a_2, \ldots$, and the values of η_2 be the numbers $b_1, b_2, \ldots$. It follows from the finiteness of the expectations $E\eta_1$ and $E\eta_2$ that

$$\sum |a_i| P\{\eta_1 = a_i\} < \infty, \quad \sum |b_i| P\{\eta_2 = b_i\} < \infty.$$

The product $\eta_1 \cdot \eta_2$ takes values $a_i b_j$ on the set $(\omega | \eta_1 = a_i) \cap (\omega | \eta_2 = b_j)$. Thus, by independence,

$$E(\eta_1 \cdot \eta_2) = \sum_i \sum_j a_i b_j P\{(\omega | \eta_1(\omega) = a_i, \eta_2(\omega) = b_j)\}$$

$$= \sum_i \sum_j a_i b_j P\{\eta_1(\omega) = a_i\} \cdot P\{\eta_2(\omega) = b_j\}$$

$$= \sum_i a_i P\{\eta_1(\omega) = a_i\} \cdot \sum_j b_j P\{\eta_2(\omega) = b_j\} = E\eta_1 \cdot E\eta_2$$

by the fact that the double series converges absolutely.

The following theorem is proved exactly in the same way.

Theorem 4.2. *Let $\eta_1, \eta_2, \ldots, \eta_n$ be jointly independent random variables such that $E\eta_i$ exists for $1 \leq i \leq n$. Then $E(\eta_1 \cdot \eta_2 \cdots \eta_n)$ exists and $E(\eta_1 \cdots \eta_n) = E\eta_1 \ldots E\eta_n$.*

Corollary. *Let $\eta_1, \ldots, \eta_n$ be pairwise independent random variables such that* Var $\eta_i < \infty$ *for* $1 \leq i \leq n$. *Then* Var$(\eta_1 + \cdots + \eta_n) < \infty$ *and* Var$(\eta_1 + \cdots \eta_n) =$ Var $\eta_1 + \cdots +$ Var η_n.

Proof. In Lecture 1 we obtained the formula

$$\mathrm{Var}(\eta_1 + \cdots + \eta_n) = \sum_{i=1}^{n} \mathrm{Var}\, \eta_i + 2 \sum_{i > j} \mathrm{Cov}(\eta_i, \eta_j).$$

But

$$\mathrm{Cov}(\eta_i, \eta_j) = E(\eta_i \eta_j) - E\eta_i \cdot E\eta_j = E\eta_i \cdot E\eta_j - E\eta_i \cdot E\eta_j = 0.$$

$\square$

4.3 The Gambler's Ruin Problem

We consider here a typical problem which can be solved by using the total probability formula. We assume that a series of games take place in which the player of interest to us wins each time with probability p and loses with probability $1 - p$ independently of the other games. We denote by x the fortune of the gambler. We will assume that when the gambler wins his fortune is increased by 1, i.e. $x \mapsto x + 1$, and when the gambler loses his fortune is decreased by 1, i.e. $x \mapsto x - 1$. The game stops when the fortune of the gambler becomes zero (gambler's ruin) or a number $a > 0$ (gambler's victory) where a is the sum of the fortunes of the gamblers at the beginning of the game. Let us denote by z the initial fortune of our gambler. The course of the game can be represented conveniently by a graph consisting of straight segments with angles of $\pm 45°$, similar to the trajectory of a random walk. Each graph starts at the point z, $0 < z < a$ and ends either at $z = a$ (victory) or at $z = 0$ (ruin). It is convenient to assume that after attaining $z = 0$ or a the graph continues as a straight horizontal line. These graphs correspond to the points ω of the space of elementary outcomes Ω. In order to indicate the dependence on the initial point z we write ω_z and Ω_z. For $z = 0$ or a we assume that Ω_0 or Ω_a contain only one element, consisting of a horizontal line. The typical form of ω is given in Fig. 4.1.

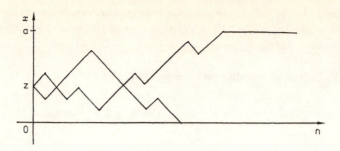

Fig. 4.1.

For $\omega_z \in \Omega_z$ we set $p(\omega_z) = p^k q^\ell$ where k (or ℓ) is the number of segments which move upwards (or downwards), i.e. with an angle $45°$ (or $-45°$). This definition does not imply that $\sum_{\omega_z \in \Omega_z} p(\omega_z) = 1$. The violation of this last equality can be interpreted as the fact that there is a positive probability for an infinite gain. Therefore we set

$$P_z = \sum_{\omega_z \in \Omega_z} p(\omega_z),\ 0 < z < a,$$
$$W_z = \Sigma' p(\omega_z),\ L_z = \Sigma'' p(\omega_z),\ 0 < z < a,$$

where Σ' (or Σ'') indicates that the summation is carried out for those ω_z that end with a (or with 0), i.e. W_z is the probability of victory and L_z is the probability of ruin with an initial fortune of z. For $z = 0$ or a it follows from the definition that:

$$P_0 = 1, W_0 = 0, L_0 = 1.$$
$$P_a = 1, W_a = 1, L_a = 0.$$

Lemma 4.2. *The following relations hold:*

$$P_z = pP_{z+1} + qP_{z-1}, 0 < z < a;$$
$$W_z = pW_{z+1} + qW_{z-1}, 0 < z < a;$$
$$L_z = pL_{z+1} + qL_{z-1}, 0 < z < a.$$

Proof. We have

$$P_z = \sum_{\omega_z \in \Omega_z} p(\omega_z) = \sum {}^{(1)} p(\omega_z) + \sum {}^{(2)} p(\omega_z),$$

where $\sum {}^{(1)}$ denotes the sum over those ω_z in which the first game was won and $\sum {}^{(2)}$ denotes the sum over those ω_z in which the first game was lost. For $\omega_z \in \sum {}^{(1)}$ we have $p(\omega_z) = p.p(\omega_{z+1})$, where ω_{z+1} is the graph which begins at $n = 1$ with the point $z + 1$. Clearly any $\omega_{z+1} \in \Omega_{z+1}$ is obtained this way. Therefore $\sum {}^{(1)} p(\omega_z) = p.\sum_{\omega_{z+1} \in \Omega_{z+1}} p(\omega_{z+1}) = p.P_{z+1}$. In the same way it is clear that for $\omega_z \in \sum^{(2)}$ we have $p(\omega_z) = q.p(\omega_{z-1})$ and therefore $\sum^{(2)} p(\omega_z) = q.P_{z-1}$. This completes the proof of the first relation. The remaining relations are proved in the same way. $\qquad\qquad\square$

In what follows we will need a general property of the solutions to the homogeneous linear equation

$$R_z = p.R_{z+1} + q.R_{z-1}, \ 0 < z < a \qquad\qquad (4.3)$$

subject to the boundary conditions $R_0 = C_1$, $R_a = C_2$.

Lemma 4.3. *There exists at most one solution to (4.3) which satisfies the given boundary conditions.*

Proof. Assume that, to the contrary, there are two such solutions, $R_z^{(1)}, R_z^{(2)}$ and let $R_z^{(3)} = R_z^{(1)} - R_z^{(2)}$. Then $R_0^{(3)} = R_a^{(3)} = 0$ and

$$R_z^{(3)} = pR_{z+1}^{(3)} + qR_{z-1}^{(3)}, \ 0 < z < a.$$

Let us choose z_0 such that $\max\{R_z^{(3)}, 0 < z < a\} = R_{z_0}^{(3)} > 0$. Since $R_{z_0}^{(3)} \geq R_{z_0+1}^{(3)}$, $R_{z_0}^{(3)} \geq R_{z_0-1}^{(3)}$ the equality

$$R_{z_0}^{(3)} = pR_{z_0+1}^{(3)} + qR_{z_0-1}^{(3)}$$

is possible if and only if $R_{z_0}^{(3)} = R_{z_0+1}^{(3)} = R_{z_0-1}^{(3)}$. In exactly the same way we can go to $z_0 \pm 2$ and so forth. In the end we obtain that $R_z^{(3)} = $ constant for $0 < z < a$. But this constant can only be 0 since $R_0^{(3)} = R_a^{(3)} = 0$. An analogous argument can be carried out in the case where $\min R_z^{(3)} < 0$. In the same way we obtain $R_z^{(3)} \equiv 0$ for $0 \leq z \leq a$, i.e. $R_z^{(1)} = R_z^{(2)}$.

The function P_z satisfies (4.3) and the boundary conditions $P_0 = P_a = 1$. Since $P_z \equiv 1$ is a solution to (4.3) with such boundary conditions, it follows from Lemma 4.2 that this solution is unique, i.e. $P_z \equiv 1$. We also have shown, in this way, that the probability of an infinite gain is equal to zero.

Given this fact we can obtain the relations of Lemma 4.1 by using the total probability formula. Let A_z be the event consisting of all ω_z which lead to victory, i.e. consisting of those graphs ω_z which end with $z = a$. Then $W_z = P(A_z)$. We denote by B_+ (or B_-) the event that the first game was won (or lost). Clearly $B_+ \cap B_- = \emptyset$, $B_+ \cup B_- = \Omega_z$. Furthermore $P(B_+) = p$, since

$$P(B_+) = \sum\nolimits^{(1)} p(\omega_z) = p. \sum_{\omega_{z+1} \in \Omega_{z+1}} p(\omega_{z+1}) = p.P_{z+1} = p.$$

Similarly

$$P(B_-) = \sum\nolimits^{(2)} p(\omega_z) = q. \sum_{\omega_{z-1} \in \Omega_{z-1}} p(\omega_{z-1}) = q.$$

By the total probability formula

$$W_z = P(B_+)P(A_z|B_+) + P(B_-)P(A_z|B_-)$$
$$= pP(A_z|B_+) + qP(A_z|B_-).$$

From the definition of conditional probability we have

$$P(A_z|B_+) = \frac{P(A_z \cap B_+)}{P(B_+)} = \frac{p.P_{z+1}}{p} = P_{z+1}.$$

Similarly $P(A_z|B_-) = P_{z-1}$. Therefore the relations of Lemma 4.1 are a particular case of the total probability formula. $\square$

We now find explicit formulas for W_z and L_z. The method is totally analogous to the method of solving linear differential equations with constant coefficients. We first seek particular solutions to (4.3) of the form $R_z = \lambda^z$. Substituting into (4.3) gives $\lambda = p\lambda^2 + q$. The roots of the last equation have the form $\lambda_1 = 1$, $\lambda_2 = q/p$.

First Case. $p \neq q$. In this case $\lambda_1 \neq \lambda_2$. We will look for W_z of the form

$$W_z = d_1\lambda_1^z + d_2\lambda_2^z = d_1 + d_2\left(\frac{q}{p}\right)^z,$$

where d_1 and d_2 will be found from the boundary conditions

$$W_0 = d_1 + d_2 = 0,$$
$$W_a = d_1 + d_2\left(\frac{q}{p}\right)^a = 1.$$

This gives $d_1 = -d_2 = 1/(1 - (q/p)^a)$. We therefore obtain

$$W_z = \frac{1 - \left(\frac{q}{p}\right)^z}{1 - \left(\frac{q}{p}\right)^a}.$$

L_z can now be found from the relation

$$L_z = 1 - W_z = \frac{\left(\frac{p}{q}\right)^{a-z} - 1}{\left(\frac{p}{q}\right)^a - 1}.$$

Second Case. $p = q = 1/2$. In this case the equation has the form

$$W_z = \frac{1}{2}(W_{z+1} + W_{z-1}),$$

and also $\lambda_1 = \lambda_2 = 1$. Aside from the particular solution $W_z = 1$ we can check immediately that $W_z = z$ is also a solution. We therefore look for a solution of the form

$$W_z = d_1 + d_2 z.$$

Taking into account the boundary conditions we obtain:

$$W_0 = d_1 = 0; \ W_a = d_2 a = 1, \ d_2 = a^{-1}.$$

Thus

$$W_z = \frac{z}{a}, \ L_z = 1 - W_z = \frac{a - z}{a}.$$

An intuitive conclusion arises from our equations: the greater the initial fortune, the greater the probability of winning.

We now introduce the random variable $\tau(w_z)$, $w_z \in \Omega_z$, equal to the duration of the game, i.e. equal to that $n = n(w_z)$ for which the graph first attains 0 or a.

Lemma 4.4. *There exist constants q, $0 < q < 1$ and c, $0 < c < \infty$, such that*

$$P\{\tau(w_z) \geq n\} \leq c.q^n.$$

Proof. Let us denote by D_n the set of graphs $\gamma_f = \{x = f(t), \ 0 \leq t \leq n\}$ which represent the course of the game from time 0 to n and for which the game at time n is not finished yet. We denote by C_f the set of those $w_z \in \Omega_z$ for which the course of the game from time 0 to n agrees with the function f. We then have

$$P_n = P\{\tau(w_z) \geq n\} = \sum_{C_f} \sum_{w_z \in C_f} p(w_z).$$

We first show that $\sum_{w_z \in C_f} p(w_z) = p^k q^{n-k}$, where k is the number of games won between the times 0 and n, i.e. the number of segments in the graph f of length $\sqrt{2}$ which move with an angle of $+45°$ (see Fig. 4.2).

Fig. 4.2.

Indeed let $f(n) = z_1$. Then every graph $w_z \in C_f$ consists of two parts: the first part describes the function f and the second part can be considered as an element w_{z_1} of the space Ω_{z_1}. Correspondingly $p(w_z) = p^k q^{n-k}.p(w_{z_1})$ and

$$\sum_{w_z \in C_f} p(w_z) = p^k q^{n-k} \sum_{w_{z_1} \in \Omega_{z_1}} p(w_{z_1}) = p^k q^{n-k}.$$

We now establish the following inequality: for some q_1, $0 < q_1 < 1$,

$$P_{n+a} \leq q_1.P_n. \tag{4.4}$$

Indeed if we denote by $f^{(1)}$ the graphs which correspond to $n + a$ and by f the graphs which correspond to n then we have

$$
P_{n+a} = \sum_{f} P(C_{f^{(1)}}) = \sum_{f} \sum_{C_{f^{(1)}} \subset C_f} P(C_{f^{(1)}})
$$

$$
= \sum_{f} \sum_{C_{f^{(1)}} \subset C_f} p^{k_1} q^{(n+a)-k_1} = \sum_{f} p^k q^{n-k} \sum_{C_{f^{(1)}} \subset C_f} p^{k_1-k} q^{a-k_1+k}.
$$

Here the notation $\sum_{C_{f^{(1)}} \subset C_f}$ denotes a summation with respect to the graphs $f^{(1)}$ which are extensions of f. We now note that

$$
\sum_{C_f^{(1)} \subset C_f} p^{k-k_1} \cdot q^{a-k_1+k} \leq q_1 < 1,
$$

since in a steps the only possible extensions $f^{(1)}$ are those which do not lead to the end of the game.

The desired statement follows from (4.4) since for any m, $m = ka + r$, $0 \leq r < a$

$$
P_m \leq q_1 \cdot P_{m-a} \leq q_1^2 \cdot P_{m-2a} \leq q_1^k \cdot P_r \leq q_1^k \leq q_1^{-1} \cdot q^m,
$$
$$
q = q_1^{1/k}.
$$

Setting $c = q_1^{-1}$, $q = q_1^{1/k}$ we obtain the statement of the lemma.   □

It follows clearly from Lemma 4.4 that for any $k > 0$

$$
E\tau^k(\omega_z) = \sum_{\omega_z \in \Omega_z} \tau^k(\omega_z) p(\omega_z) < \infty.
$$

Set $E_z = E\tau(\omega_z)$.

Lemma 4.5. $E_z = pE_{z+1} + qE_{z-1} + 1$, $0 < z < a$, $E_0 = E_a = 0$.

Proof. The last equalities follow from the definition. To prove the first equality we write

$$
E_z = \sum_z \tau(\omega_z) p(\omega_z) = \sum_{\omega_z \in B_+} \tau(\omega_z) p(\omega_z) + \sum_{\omega_z \in B_-} \tau(\omega_z) p(\omega_z)
$$

$$
= \sum^{(1)}_{\omega_{z+1} \in \Omega_{z+1}} (\tau(\omega_{z+1}) + 1) \cdot p \cdot p(\omega_{z+1})
$$

$$
+ \sum^{(2)}_{\omega_{z-1} \in \Omega_{z-1}} (\tau(\omega_{z-1}) + 1) q \cdot p(\omega_{z-1})
$$

$$
= pE_{z+1} + p + qE_{z-1} + q = pE_{z+1} + qE_{z-1} + 1.
$$

Here we used the fact that in the sum $\sum^{(1)}$ (or $\sum^{(2)}$) each ω_z can be represented as a union of the first segment moving with an angle $+45°$ (or $-45°$) and of $\omega_{z+1} \in \Omega_{z+1}$ (or $\omega_{z-1} \in \Omega_{z-1}$), and that each ω_{z+1} (or ω_{z-1}) appears in such a representation.   □

The statement of Lemma 4.3 also applies to E_z. Therefore if we find one function E_z which satisfies the conditions in Lemma 4.5 that function will be the one we are looking for.

First case. $p \neq q$. We note that the function $E_z = z/(q-p)$ satisfies the equation

$$E_z = pE_{z+1} + qE_{z-1} + 1, \tag{4.5}$$

but does not satisfy the boundary conditions. Since every function of the form $R_z = d_1 + d_2 (q/p)^z$ satisfies the equation $R_z = pR_{z+1} + qR_{z-1}$ then any function $E_z = z/(q-p) + R_z$ satisfies (4.5). We choose d_1 and d_2 in order to satisfy the boundary conditions

$$E_0 = d_1 + d_2 = 0, \; d_2 = -d_1$$

$$E_a = \frac{a}{q-p} + d_1 \left(1 - \left(\tfrac{q}{p}\right)^a\right) = 0, \; d_1 = \frac{a}{(p-q)\left(1 - \left(\tfrac{q}{p}\right)^a\right)}.$$

Finally we obtain

$$E_z = \frac{z}{q-p} + \frac{a\left(1 - \left(\tfrac{q}{p}\right)^z\right)}{(p-q)\left(1 - \left(\tfrac{q}{p}\right)^a\right)} = \frac{a - z - \left(a\left(\tfrac{q}{p}\right)^z + z\left(\tfrac{q}{p}\right)^a\right)}{(p-q)\left(1 - \left(\tfrac{q}{p}\right)^a\right)}.$$

Second Case. $p = q = 1/2$. The equation (4.5) then has the form

$$E_z = \frac{1}{2}(E_{z+1} + E_{z-1}) + 1. \tag{4.6}$$

The function $E_z = -z^2$ satisfies (4.6). As in the first case we look for E_z of the form:

$$E_z = -z^2 + d_1 + d_2 z.$$

Then

$$E_0 = d_1 = 0,$$
$$E_z = -a^2 + d_2 a = 0, \; d_2 = a.$$

Therefore

$$E_z = -z^2 + az = z(a - z).$$

The function E_z attains its maximum value $a^2/4$ for $z = a/2$, which is natural in view of the symmetry $p = q = 1/2$. In addition this relation shows that the duration of the game for $z \sim a/2$ is of the order $a^2/4$, i.e. of the order of the square of the length of the interval $[0, a]$. We have already seen that such a behavior is typical for random walks.

Lecture 5. Markov Chains

5.1 Stochastic Matrices

The theory of Markov chains makes use of the theory of so-called stochastic matrices. We therefore begin with a small digression of a purely algebraic nature.

Definition 5.1. A matrix $Q = \|q_{ij}\|$ of order $r \times r$ is said to be stochastic if

1. $q_{ij} \geq 0$,
2. $\sum_{j=1}^{r} q_{ij} = 1$, for any i.

The conditions under which a matrix is stochastic can be expressed somewhat differently. A column vector $\vec{f} = (f_1, \ldots, f_r)'$ is said to be non-negative if $f_i \geq 0$, $1 \leq i \leq r$; in this case we write $\vec{f} \geq 0$.

Lemma 5.1. *The following statements, a), b.1), b.2) and c), are equivalent:*

a) *the matrix Q is stochastic;*

b.1) *for any $\vec{f} \geq 0$ the vector $Q\vec{f} \geq 0$;*

b.2) *if $\mathbf{1} = (1, \ldots, 1)'$ then $Q\mathbf{1} = \mathbf{1}$, i.e. the vector $\mathbf{1}$ is an eigenvector of the matrix Q corresponding to the eigenvalue 1;*

c) *if $\vec{\mu} = (\mu_1, \ldots, \mu_r)$ is a probability distribution, i.e. $\mu_i \geq 0$, $\sum_{i=1}^{r} \mu_i = 1$ then $\vec{\mu}' = \vec{\mu}Q$ is also a probability distribution.*

Proof. If Q is a stochastic matrix then b.1) and b.2) clearly hold and therefore a) $\Rightarrow$ b). We now show that b) $\Rightarrow$ a). Consider the column vector $\vec{\delta_j} = (0, \ldots, \underbrace{0, 1, 0, \ldots}_{j-1})' \geq 0$. Then $Q\vec{\delta_j} = (q_{1j}, q_{2j}, \ldots, q_{rj})' \geq 0$, i.e. $q_{ij} \geq 0$. Furthermore $Q\mathbf{1} = (\sum_j q_{1j}, \sum_j q_{2j}, \ldots, \sum_j q_{rj})'$ and it follows from the equality $Q\mathbf{1} = \mathbf{1}$ that $\sum_j q_{ij} = 1$ for all i, therefore b) $\Rightarrow$ a).

We now show that a) $\Rightarrow$ c). If $\vec{\mu}' = \vec{\mu}Q$, then $\mu'_j = \sum_{i=1}^{r} \mu_i q_{ij}$. Since Q is stochastic we have $\mu'_j \geq 0$ and $\sum_{j=1}^{r} \mu'_j = \sum_j \sum_i \mu_i q_{ij} = \sum_i \sum_j \mu_i q_{ij} = \sum_i \mu_i \sum_j q_{ij} = \sum_i \mu_i = 1$, therefore $\vec{\mu}'$ is also a probability distribution.

Now assume that c) holds. Consider the row vector $\vec{\delta_i} = (\underbrace{0,\ldots,0,1,0,\ldots,0}_{i-1})$,

which corresponds to the probability distribution on the set $(1,2,\ldots,r)$ which is concentrated at the point i. Then $\vec{\delta_i}\,Q = (q_{i1}, q_{i2}, \ldots, q_{ir})$ is also a probability distribution. It follows that $q_{ij} \geq 0$ and $\sum_{j=1}^{r} q_{ij} = 1$, i.e. c) $\Rightarrow$ a). $\square$

Lemma 5.2. *If* $Q' = \|q'_{ij}\|$, $Q'' = \|q''_{ij}\|$ *are stochastic matrices, then* $Q''' = Q'.Q''$ *is also a stochastic matrix. If all the* $q''_{ij} > 0$ *then all the* $q'''_{ij} > 0$.

Proof. We have

$$q'''_{ij} = \sum_{k=1}^{r} q'_{ik} \cdot q''_{kj}.$$

Therefore $q'''_{ij} \geq 0$. If all $q''_{kj} > 0$, then $q'''_{ij} > 0$ since $q'_{ik} \geq 0$ and $\sum_{k=1}^{r} q'_{ik} = 1$. Furthermore

$$\sum_{j=1}^{r} q'''_{ij} = \sum_{j=1}^{r}\sum_{k=1}^{r} q'_{ik} q''_{kj} = \sum_{k=1}^{r} q'_{ik} \sum_{j=1}^{r} q''_{kj} = \sum_{k=1}^{r} q'_{ik} = 1.$$

This completes the proof of the lemma. $\square$

5.2 Markov Chains

We now return to probability theory. Let Ω be a space of elementary outcomes $\omega = \{\omega_0, \omega_1, \ldots, \omega_n\}$, where $\omega_i \in X = \{x^{(1)}, x^{(2)}, \ldots, x^{(r)}\}$, $0 \leq i \leq n$. Moreover let us call ω_0 the state at the origin of time, and ω_i the state at time i. We assume, as given, a probability distribution $\mu = \{\mu_1, \ldots, \mu_r\}$ on X and n stochastic matrices $P(1), \ldots, P(n)$, with $P(k) = \|p_{ij}(k)\|$.

Definition 5.2. A Markov Chain with state space X, generated by the initial distribution $\vec{\mu}$ on X and the stochastic matrices $P(1), \ldots, P(n)$, is a probability distribution P on Ω which satisfies

$$p(\omega) = \mu_{\omega_0} \cdot p_{\omega_0\omega_1}(1) \cdot p_{\omega_1\omega_2}(2) \cdot \ldots \cdot p_{\omega_{n-1}\omega_n}(n).$$

The points $x^{(j)} \in X$, $1 \leq j \leq r$, are called the states of the Markov chain.

We now check that the last equation defines a probability distribution on Ω. The inequality $p(\omega) \geq 0$ is clear. It remains to be shown that $\sum_{\omega \in \Omega} p(\omega) = 1$. We have

$$\Sigma = \sum_{\omega \in \Omega} p(\omega) = \sum_{\omega_0,\ldots,\omega_n} \mu_{\omega_0} \cdot p_{\omega_0\omega_1}(1) \cdot \ldots \cdot p_{\omega_{n-1}\omega_n}(n).$$

We now perform the summation as follows: Fix the values $\omega_0, \omega_1, \ldots, \omega_{n-1}$ and sum over all values of ω_n. Thus

$$\Sigma = \sum_{\omega_0, \dots, \omega_{n-1}} \mu_{\omega_0} \cdot p_{\omega_0 \omega_1}(1) \cdot \dots \cdot p_{\omega_{n-2}\omega_{n-1}}(n-1) \cdot \sum_{\omega_n = 1}^{r} p_{\omega_{n-1}\omega_n}(n).$$

The last sum is equal to 1 by virtue of the fact that $P(n)$ is a stochastic matrix. The remaining sum has the same form as Σ, where n has been replaced by $n-1$. We then fix $\omega_0, \dots, \omega_{n-2}$ and sum over all values of ω_{n-1}, and so on. In the end we obtain $\sum_{\omega_0=1}^{r} \mu_{\omega_0} = 1$ by the fact that $\vec{\mu}$ is a probability distribution. It follows that $\Sigma = 1$.

By similar arguments one can prove the following statement:

$$P\{\omega_0 = x^{(i_0)}, \omega_1 = x^{(i_1)}, \dots, \omega_k = x^{(i_k)}\} = \mu_{i_0} \cdot p_{i_0 i_1}(1) \cdot \dots \cdot p_{i_{k-1} i_k}(k)$$

for any $x^{(i_0)}, \dots, x^{(i_k)}, k \le n$. This equality shows that the induced probability distribution on the space of $(k+1)$-tuples $(\omega_0, \dots, \omega_k)$ is also a Markov chain, generated by the initial distribution $\vec{\mu}$, and the stochastic matrices $P(1), \dots, P(k)$.

The matrix entries $p_{ij}(k)$ are called the transition probabilities from state $x^{(i)}$ to state $x^{(j)}$ at time k.

Assume that $P\{\omega_0 = x^{(i_0)}, \dots, \omega_{k-2} = x^{(i_{k-2})}, \omega_{k-1} = x^{(i)}\} > 0$. We consider the conditional probability $P\{\omega_k = x^{(j)} | \omega_{k-1} = x^{(i)}, \omega_{k-2} = x^{(i_{k-2})}, \dots, \omega_0 = x^{(i_0)}\}$. By definition:

$$P\{\omega_k = x^{(j)} | \omega_{k-1} = x^{(i)}, \omega_{k-2} = x^{(i_{k-2})}, \dots, \omega_0 = x^{(i_0)}\}$$

$$= \frac{P\{\omega_0 = x^{(i_0)}, \dots, \omega_{k-2} = x^{(i_{k-2})}, \omega_{k-1} = x^{(i)}, \omega_k = x^{(j)}\}}{P\{\omega_0 = x^{(i_0)}, \dots, \omega_{k-2} = x^{(i_{k-2})}, \omega_{k-1} = x^{(i)}\}}$$

$$= \frac{\mu_{i_0} \cdot p_{i_0 i_1}(1) \cdot p_{i_1 i_2}(2) \cdot \dots \cdot p_{i_{k-2} i}(k-1) \cdot p_{ij}(k)}{\mu_{i_0} \cdot p_{i_0 i_1}(1) \cdot p_{i_1 i_2}(2) \cdot \dots \cdot p_{i_{k-2} i}(k-1)}$$

$$= p_{ij}(k)$$

and does not depend on the values $i_0, i_1, \dots, i_{k-2}$. This property is sometimes used as the definition of a Markov chain.

Definition 5.3. A Markov chain is said to be homogeneous if $P(k) = P$ does not depend on k, $1 \le k \le n$.

Homogeneous Markov chains can be understood as a generalization of sequences of independent identical trials. Indeed if the stochastic matrix $P = \|p_{ij}\|$ is such that all its rows are equal to $\{p_1, \dots, p_r\}$, where $\{p_1, \dots, p_r\}$ is a probability distribution on X, then a Markov chain with such a matrix P is a sequence of independent identical trials.

In what follows we consider only homogeneous Markov chains. Sometimes we will represent such chains with a graph. The vertices of the graph will consist of the points $x^{(i)} \in X$. The points $x^{(i)}$ and $x^{(j)}$ are connected by an oriented edge if $p_{ij} > 0$. A collection of states $\omega_0, \omega_1, \dots, \omega_k$, which has a positive probability, can be represented on the graph as a path of length k starting at the point ω_0; $P(\omega_0, \omega_1, \dots, \omega_k)$ is the probability of such a path.

Therefore a homogeneous Markov chain can be represented as a probability distribution on the space of paths of length k on the graph, or as a sequence of jumps across the vertices of the graph.

We consider the conditional probabilities $P(\omega_{s+\ell} = x^{(j)}|\omega_\ell = x^{(i)})$. It is clearly assumed here that $P(\omega_\ell = x^{(i)}) > 0$. We show that these conditional probabilities do not depend on ℓ and that $P(\omega_{s+\ell} = x^{(j)}|\omega_\ell = x^{(i)}) = p_{ij}^{(s)}$, where the $p_{ij}^{(s)}$ are the elements of the matrix P^s. By Lemma 5.2 the matrix P^s is stochastic.

For $s = 1$, using the homogeneity of the chain, we have

$$
P(\omega_{\ell+1} = x^{(j)}|\omega_\ell = x^{(i)}) = \frac{P(\omega_{\ell+1} = x^{(j)}, \omega_\ell = x^{(i)})}{P(\omega_\ell = x^{(i)})}
$$

$$
= \frac{\displaystyle\sum_{x^{(i_0)},\dots,x^{(i_{\ell-1})}} P(\omega_0 = x^{(i_0)},\dots,\omega_{\ell-1} = x^{(i_{\ell-1})},\omega_\ell = x^{(i)},\omega_{\ell+1} = x^{(j)})}{\displaystyle\sum_{x^{(i_0)},\dots,x^{(i_{\ell-1})}} P(\omega_0 = x^{(i_0)},\dots,\omega_\ell = x^{(i)})}
$$

$$
= \frac{\displaystyle\sum_{i_0,\dots,i_{\ell-1}} \mu_{i_0}\cdot p_{i_0 i_1}\cdot p_{i_1 i_2}\cdots p_{i_{\ell-1} i} p_{ij}}{\displaystyle\sum_{i_0,\dots,i_{\ell-1}} \mu_{i_0}\cdot p_{i_0 i_1}\cdot p_{i_1 i_2}\cdots p_{i_{\ell-1} i}} = p_{ij},
$$

i.e. we have proved our statement for $s = 1$. Let us assume that the statement is proved for all $s \le s_0$ and let us prove it for $s_0 + 1$. We have

$$
P(\omega_{s_0+1+\ell} = x^{(j)}|\omega_\ell = x^{(i)}) = \frac{P(\omega_{s_0+1+\ell} = x^{(j)}, \omega_\ell = x^{(i)})}{P(\omega_\ell = x^{(i)})}
$$

$$
= \frac{\displaystyle\sum_{x^{(k)}} P(\omega_{s_0+1+\ell} = x^{(j)}, \omega_{s_0+\ell} = x^{(k)}, \omega_\ell = x^{(i)})}{P(\omega_\ell = x^{(i)})}
$$

$$
= \sum_{x^{(k)}} \frac{P(\omega_{s_0+1+\ell} = x^{(j)}|\omega_{s_0+\ell} = x^{(k)}, \omega_\ell = x^{(i)})}{P(\omega_\ell = x^{(i)})}
$$

$$
\times P(\omega_{s_0+\ell} = x^{(k)}, \omega_\ell = x^{(i)})
$$

$$
= \sum_{k=1}^{r} P(\omega_{s_0+1+\ell} = x^{(j)}|\omega_{s_0+\ell} = x^{(k)}, \omega_\ell = x^{(i)})
$$

$$
\times P(\omega_{s_0+\ell} = x^{(k)}|\omega_\ell = x^{(i)}).
$$

By the induction hypothesis the last factor is equal to $p_{ik}^{(s_0)}$. We show that the first factor is equal to p_{kj}. Indeed we have

$$P\big(\omega_{s_0+1+\ell} = x^{(j)} \big| \omega_{s_0+\ell} = x^{(k)}, \omega_\ell = x^{(i)}\big)$$

$$= \frac{P\big(\omega_{s_0+1+\ell} = x^{(j)}, \omega_{s_0+\ell} = x^{(k)}, \omega_\ell = x^{(i)}\big)}{P\big(\omega_{s_0+\ell} = x^{(k)}, \omega_\ell = x^{(i)}\big)}$$

$$= \frac{1}{P\big(\omega_{x_0+\ell} = x^{(k)}, \omega_\ell = x^{(i)}\big)} \sum_{i_{s_0+\ell-1},\dots,i_0} P\big(\omega_{s_0+1+\ell} = s^{(j)},$$

$$\omega_{s_0+\ell} = x^{(k)}, \omega_{s_0+\ell-1} = x^{(i_{s_0+\ell-1})},\dots,\omega_{\ell+1} = x^{(i_{\ell+1})},$$

$$\omega_\ell = x^{(i)}, \omega_{\ell-1} = x^{(i_{\ell-1})},\dots,\omega_0 = x^{(i_0)}\big)$$

$$= \frac{1}{P\big(\omega_{s_0+\ell} = x^{(k)}, \omega_\ell = x^{(i)}\big)} \sum_{i_0,\dots,i_{\ell-1},i_{\ell+1},\dots,i_{s_0+\ell-1}} \mu_{i_0}$$

$$p_{i_0 i_1} \cdots p_{i_{\ell-1} i} p_{i i_{\ell+1}} \cdots p_{i_{s_0+\ell-1} k} p_{kj}$$

$$= \frac{p_{kj}}{P\big(\omega_{s_0+\ell} = x^{(k)}, \omega_\ell = x^{(i)}\big)}$$

$$\times \sum_{\substack{i_0,\dots,i_{\ell-1},\\ i_{\ell+1},\dots,i_{s_0+\ell-1}}} \mu_{i_0} p_{i_0 i_1} \cdots p_{i_{\ell-1} i} p_{i i_{\ell+1}} \cdots p_{i_{s_0+\ell-1} k} = p_{kj}.$$

By the same token

$$P\big(\omega_{s_0+\ell+1} = x^{(j)} \big| \omega_{s_0+\ell} = x^{(k)}, \omega_\ell = x^{(i)}\big) = \sum_{k=1}^{r} p_{ik}^{(s_0)} \cdot p_{kj} = p_{ij}^{(s_0+1)}.$$

This completes the proof of our statement.

The transition probabilities $p_{ij}^{(s)}$ are called s—step transition probabilities.

Definition 5.4. The matrix P is said to be ergodic if there exists an s_0 such that $p_{ij}^{(s_0)} > 0$ for any i, j.

By the statement we have just proved this means that in s_0 steps one can, with positive probability, proceed from any initial state $x^{(i)}$ to any final state $x^{(j)}$.

5.3 Non-Ergodic Markov Chains

We now construct examples of Markov chains by giving their graphs. So we draw graphs without ergodicity.

In the periodic graph as presented in the figure the states are divided into three pairs, and each time a transition occurs from one pair to another. For any given s the transition probability $p_{ij}^{(s)} \neq 0$ only for two appropriate values of j. It is clear that periodic graphs are possible with any period.

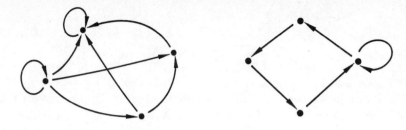

Fig. 5.1. Disconnected graph

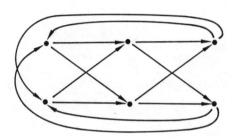

Fig. 5.2. Periodic graph

One can show that the examples we have given, and combinations of them, describe all examples of non-ergodic Markov chains.

Theorem 5.1 (Ergodic Theorem for Markov Chains). *Assume as given a Markov chain with an ergodic transition probability matrix P. Then there exists a unique probability distribution $\pi = (\pi_1, \ldots, \pi_r)$ such that*

1. $\pi P = \pi$,

2. $\lim_{s \to \infty} p_{ij}^{(s)} = \pi_j$.

Proof. Let $\mu' = \{\mu'_1, \ldots, \mu'_r\}$, $\mu'' = \{\mu''_1, \ldots, \mu''_r\}$ be two probability distributions on the space X. We set $d(\mu', \mu'') = (1/2) \sum_{i=1}^r |\mu'_i - \mu''_i|$. Then d can be considered as a distance on the space of probability distributions on X, and the space X with that distance is a complete metric space. We note now that

$$0 = \sum_{i=1}^r \mu'_i - \sum_{i=1}^r \mu''_i = \sum_{i=1}^r (\mu'_i - \mu''_i) = \sum{}^{+} (\mu'_i - \mu''_i) - \sum{}^{+} (\mu'_i - \mu''_i),$$

where $\sum^{+}$ from now on will denote summation with respect to those indices i for which the terms are positive. Therefore

$$d(\mu',\mu'') = \frac{1}{2}\sum_{i=1}^{r}|\mu_i' - \mu_i''| = \frac{1}{2}\sum\nolimits^{+}(\mu_i' - \mu_i'')$$

$$+ \frac{1}{2}\sum\nolimits^{+}(\mu_i'' - \mu_i') = \sum\nolimits^{+}(\mu_i' - \mu_i'').$$

We will soon use this formula. It is also clear that $d(\mu',\mu'') \leq 1$.

Let μ' and μ'' be two probability distributions on X. By Lemma 5.1 $\mu'Q$ and $\mu''Q$ are also probability distributions on X, for any stochastic matrix Q.

Lemma 5.3.

a. $d(\mu'Q,\mu''Q) \leq d(\mu',\mu'')$;
b. if all $q_{i,j} \geq \alpha$ then $d(\mu'Q,\mu''Q) \leq (1-\alpha).d(\mu',\mu'')$.

Proof. We have

$$(\mu'Q)_j = \sum_i \mu_i' q_{ij}, \qquad (\mu''Q)_j = \sum_i \mu_i'' q_{ij}.$$

Furthermore

$$d(\mu'Q,\mu''Q) = \sum\nolimits_j^{+}\sum\nolimits_i (\mu_i' - \mu_i'')q_{ij}$$

$$\leq \sum\nolimits_j^{+}\sum\nolimits_i^{+}(\mu_i' - \mu_i'')q_{ij} = \sum\nolimits_i^{+}(\mu_i' - \mu_i'')\sum\nolimits_j^{+}q_{ij}.$$

The sum $\sum_j^{+} q_{ij} \leq \sum_j q_{ij} = 1$ and this completes the proof of a). We now note that the sum $\sum_j^{+}$ cannot be a sum over all indices j. Indeed if $\sum_{i=1}^{r}\mu_i' q_{ij} > \sum_{i=1}^{r}\mu_i'' q_{ij}$ for all j, then

$$\sum_{j=1}^{r}\sum_{i=1}^{r}\mu_i' q_{ij} > \sum_{j=1}^{r}\sum_{i=1}^{r}\mu_i'' q_{ij},$$

which is impossible, since each of the sums is equal to 1. This latter fact is easily seen by interchanging the order of summation. Therefore at least one index j is missing in the sum $\sum_j^{+} q_{ij}$. Thus if all $q_{ij} > \alpha$ then $\sum_j^{+} q_{ij} < 1 - \alpha$ and $d(\mu'Q,\mu''Q) \leq (1-\alpha)\sum_i^{+}(\mu_i' - \mu_i'') = (1-\alpha)d(\mu',\mu'')$.   □

Let μ_0 be an arbitrary probability distribution on X and let $\mu_n = \mu_0 P^n$. We show that the sequence of probability distributions μ_n is a Cauchy sequence. This means that for any $\epsilon > 0$ there exists an $n_0(\epsilon)$ such that for any p we have $d(\mu_n,\mu_{n+p}) < \epsilon$ for $n \geq n_0(\epsilon)$. By Lemma 5.3 we have

$$d(\mu_n,\mu_{n+p}) = d(\mu_0 P^n, \mu_0 P^{n+p})$$

$$\leq (1-\alpha)d(\mu_0 P^{n-s_0}, \mu_0 P^{n+p-s_0})$$

$$\leq (1-\alpha)^2 d(\mu_0 P^{n-2s_0}, \mu_0 P^{n+p-2s_0}) \leq \ldots$$

$$\leq (1-\alpha)^m d(\mu_0 P^{n-ms_0}, \mu_0 P^{n+p-ms_0})$$

$$\leq (1-\alpha)^m,$$

where $n = ms_0 + s_1, 0 \le s_1 < s_0$. For sufficiently large n we have $(1-\alpha)^m \le \epsilon_0$, which completes the proof.

So the sequence $\mu_n = \mu_0 P^n$ is a Cauchy sequence. Let us set $\pi = \lim_{n \to \infty} \mu_n$. We have $\pi P = \lim_{n \to \infty} \mu_n P = \lim_{n \to \infty} (\mu_0 P^n) P = \lim_{n \to \infty} (\mu_0 P^{n+1}) = \pi$, i.e. $\pi P = \pi$. We now show that a vector π with such properties is unique. Let there be two vectors $\pi_1 = \pi_1.P$, $\pi_2 = \pi_2 P$. Then $\pi_1 = \pi_1 P^{s_0}$, $\pi_2 = \pi_2 P^{s_0}$. Therefore $d(\pi_1, \pi_2) = d(\pi_1 P^{s_0}, \pi_2 P^{s_0}) \le (1-\alpha) d(\pi_1, \pi_2)$ by the lemma. It follows that $d(\pi_1, \pi_2) = 0$, i.e. $\pi_1 = \pi_2$.

We have thus obtained that for any initial distribution μ_0 the limit $\lim_{n \to \infty} \mu_0 P^n = \pi$ exists and does not depend on the choice of μ_0. Let us choose, for μ_0 the probability distribution which is concentrated at the point $x^{(i)}$, i.e.

$$\mu_0 = \{\underbrace{0, \ldots, 0}_{i-1}, 1, 0, \ldots 0\}.$$

Then $\mu_0 P^n$ is the probability distribution $\{p_{ij}^{(n)}\}$. Therefore $\lim_{n \to \infty} p_{ij}^{(n)} = \pi_j$. This completes the proof of the theorem.   □

Remark. Let μ_0 be concentrated at the point $x^{(i)}$. Then $d(\mu_0 P^n, \pi) = d(\mu_0 P^n, \pi P^n) \le \ldots \le (1-\alpha)^m d(\mu_0 P^{n-m s_0}, \pi P^{n-m s_0}) \le (1-\alpha)^m$. In this relation we can take $m \ge (n/s_0) - 1$. Therefore

$$\begin{aligned} d(\mu_0 P^n, \pi) &\le (1-\alpha)^{n/s_0 - 1} \\ &\le (1-\alpha)^{-1} ((1-\alpha)^{1/s_0})^n \\ &= (1-\alpha)^{-1} \beta^n, \ \beta = (1-\alpha)^{1/s_0} < 1. \end{aligned}$$

In other words, the speed of convergence of $p_{ij}^{(n)}$ to the limit π_j is exponential.

Remark. The term "ergodicity" is borrowed from statistical mechanics. In our case it implies that in Markov chains a certain "forgetting" of initial conditions occurs, as the probability distribution of the states of the system at time n as $n \to \infty$ becomes independent of the initial distribution.

Definition 5.5. A probability distribution π for which $\pi = \pi P$ is said to be a stationary distribution of the Markov chain.

We now elucidate the probabilistic meaning of the probabilities π_i.

5.4 The Law of Large Numbers and the Entropy of a Markov Chain

As in the case of a sequence of independent identical trials we introduce the random variable $\nu^{(i)}(\omega)$ equal to the number of occurrences of the state $x^{(i)}$ in the sequence $\omega = \{\omega_0, \ldots, \omega_n\}$, i.e. the number of those k's for which

$\omega_k = x^{(i)}$. We also introduce the random variables $\nu^{(ij)}(\omega)$, equal to the number of those $k > 0$ such that $\omega_{k-1} = x^{(i)}$, $\omega_k = x^{(j)}$.

Theorem 5.2. *As $n \to \infty$ for any $\epsilon > 0$ we have*

$$P\{\omega| \left|\frac{\nu^{(i)}(\omega)}{n} - \pi_i\right| \geq \epsilon\} \to 0, \qquad \text{for} \quad 1 \leq i \leq r,$$

$$P\{\omega| \left|\frac{\nu^{(ij)}(\omega)}{n} - \pi_i p_{ij}\right| \geq \epsilon\} \to 0, \qquad \text{for} \quad 1 \leq i,j \leq r.$$

Proof. As previously, we set

$$\chi_k^{(i)}(\omega) = \begin{cases} 1, & \omega_k = x^{(i)}, \\ 0, & \omega_k \neq x^{(i)}, \end{cases}$$

$$\chi_k^{(ij)}(\omega) = \begin{cases} 1, & \omega_{k-1} = x^{(i)}, \omega_k = x^{(j)}; \\ 0, & \text{otherwise}. \end{cases}$$

Then

$$\nu^{(i)}(\omega) = \sum_{k=0}^{n} \chi_k^{(i)}(\omega), \quad \nu^{(ij)}(\omega) = \sum_{k=0}^{n} \chi_k^{(ij)}(\omega).$$

For an initial distribution $\{\mu_i\}$ we have

$$E\chi_k^{(i)}(\omega) = \sum_{m=1}^{r} \mu_m p_{mi}^{(k)}, \qquad E\chi_k^{(ij)}(\omega) = \sum_{m=1}^{r} \mu_m p_{mi}^{(k)} p_{ij}.$$

As $k \to \infty$, $p_{mi}^{(k)} \to \pi_i$. Therefore as $k \to \infty$

$$E\chi_k^{(i)}(\omega) \to \pi_i, \qquad E\chi_k^{(ij)}(\omega) \to \pi_i p_{ij},$$

exponentially fast. Consequently

$$E\left[\frac{\nu^{(i)}(\omega)}{n}\right] = E\left[\frac{1}{n}\sum_{k=1}^{n} \chi_k^{(i)}(\omega)\right] \to \pi_i, \qquad E\left[\frac{1}{n}\sum_{k=1}^{n} \chi_k^{(ij)}(\omega)\right] \to \pi_i p_{ij}.$$

Therefore for sufficiently large n we have

$$\{\omega| \left|\frac{\nu^{(i)}(\omega)}{n} - \pi_i\right| \geq \epsilon\} \subseteq \{\omega| \left|\frac{\nu^{(i)}(\omega)}{n} - \frac{1}{n}E\nu^{(i)}\right| \geq \frac{\epsilon}{2}\},$$

$$\{\omega| \left|\frac{\nu^{(ij)}(\omega)}{n} - \pi_i p_{ij}\right| \geq \epsilon\} \subseteq \{\omega| \left|\frac{\nu^{(ij)}(\omega)}{n} - \frac{1}{n}E\nu^{(ij)}\right| \geq \frac{\epsilon}{2}\}.$$

The probabilities of the events on the right hand side can be estimated by Chebyshev's inequality:

$$P\{\omega| \left|\frac{\nu^{(i)}(\omega)}{n} - \frac{1}{n}E\nu^{(i)}\right| \geq \frac{\epsilon}{2}\} = P\{\omega| \left|\nu^{(i)}(\omega) - E\nu^{(i)}\right| \geq \frac{\epsilon n}{2}\}$$

$$\leq \frac{4\mathrm{Var}\,\nu^{(i)}}{\epsilon^2 n^2};$$

$$P\{\omega| \left|\frac{\nu^{(ij)}(\omega)}{n} - \frac{1}{n}E\nu^{(ij)}\right| \geq \frac{\epsilon}{2}\} = P\{\omega| \left|\nu^{(ij)}(\omega) - E\nu^{(ij)}\right| \geq \frac{\epsilon n}{2}\}$$

$$\leq \frac{4\operatorname{Var}\nu^{(ij)}}{\epsilon^2 n^2}.$$

So the matter is reduced to the estimation of $\operatorname{Var}\nu^{(i)}$ and $\operatorname{Var}\nu^{(ij)}$. If we set $m_k^{(i)} = E\chi_k^{(i)} = \sum_{s=1}^r \mu_s \cdot p_{si}^{(k)}$, we have

$$\operatorname{Var}\nu^{(i)} = E\left(\sum_{k=1}^n (\chi_k^{(i)} - m_k^{(i)})\right)^2$$

$$= \sum_{k=1}^n E(\chi_k^{(i)} - m_k^{(i)})^2 + 2\sum_{k_1 < k_2} E(\chi_{k_1}^{(i)} - m_{k_1}^{(i)})(\chi_{k_2}^{(i)} - m_{k_2}^{(i)}).$$

Since $0 \leq \chi_k^{(i)} \leq 1$, we have $-1 \leq \chi_k^{(i)} - m_k^{(i)} \leq 1$, $(\chi_k^{(i)} - m_k^{(i)})^2 \leq 1$ and $\sum_{k=1}^n E(\chi_k^{(i)} - m_k^{(i)})^2 \leq n$. Furthermore

$$E(\chi_{k_1}^{(i)} - m_{k_1}^{(i)})(\chi_{k_2}^{(i)} - m_{k_2}^{(i)}) = E\chi_{k_1}^{(i)}\chi_{k_2}^{(i)} - m_{k_1}^{(i)}m_{k_2}^{(i)}$$

$$= \sum_{s=1}^r \mu_s p_{si}^{(k_1)} p_{ii}^{(k_2-k_1)} - m_{k_1}^{(i)}m_{k_2}^{(i)} = R_{k_1,k_2}.$$

By the ergodic theorem

$$m_k^{(i)} = \pi_i + d_k^{(i)}, \qquad |d_k^{(i)}| \leq c\lambda^k,$$

$$p_{si}^{(k)} = \pi_i + \beta_{k,s}^{(i)}, \qquad |\beta_{k,s}^{(i)}| \leq c\lambda^k$$

for some constants $c < \infty$ and $\lambda < 1$. Therefore

$$|R_{k_1,k_2}| = \left|\sum \mu_s (\pi_i + \beta_{k_1,s}^{(i)})(\pi_i + \beta_{k_2-k_1,i}^{(i)}) - (\pi_i + d_{k_1}^{(i)})(\pi_i + d_{k_2}^{(i)})\right|$$

$$\leq |\beta_{k_1,s}^{(i)}| + |\beta_{k_2-k_1,s}^{(i)}| + |d_{k_1}^{(i)}| + |d_{k_2}^{(i)}|$$

$$\leq c\lambda^{k_1} + c\lambda^{k_2-k_1} + c\lambda^{k_1} + c\lambda^{k_2}.$$

So $\sum_{k_1 < k_2} R_{k_1,k_2} \leq \text{const} \cdot n$ and consequently $\operatorname{Var}\nu^{(i)} \leq \text{const} \cdot n$. The variance $\operatorname{Var}\nu^{(ij)}$ is estimated in an analogous way. $\square$

We now draw a conclusion from this theorem on the entropy of a Markov chain. In the case of a sequence of independent identical trials we have seen that the entropy is equal to the limit as $n \to \infty$ of $-(1/n)\ln p(\omega)$ for typical ω's, i.e. for ω's which lie in a set with a large probability. In order to use this property to derive a general definition of entropy we need to study the behavior of $\ln p(\omega)$ for typical ω's in the case of a Markov chain. We have

$$p(\omega) = \mu_{\omega_0} \cdot \prod_{i,j} p_{ij}^{\nu^{(ij)}(\omega)}$$

$$= e^{\ln \mu_{\omega_0} + \sum_{i,j} \nu^{(ij)}(\omega) \ln p_{ij}},$$

$$\ln p(\omega) = \ln \mu_{\omega_0} + \sum_{i,j} \nu^{(ij)}(\omega) \ln p_{ij}.$$

From the law of large numbers, for typical ω's we have $\nu^{(ij)}(\omega)/n \sim \pi_i p_{ij}$. Therefore for such ω's we have

$$-\frac{1}{n}\ln p(\omega) = -\frac{1}{n}\ln \mu_{\omega_0} - \sum \frac{\nu^{(ij)}(\omega)}{n}\ln p_{ij} \sim -\sum_{i,j}\pi_i p_{ij}\ln p_{ij}.$$

So it is natural in the case of a Markov chain to define the entropy to be equal to $h = -\sum_i \pi_i \sum_j p_{ij}\ln p_{ij}$. It is not difficult to show that for such a definition of h Macmillan's Theorem remains true.

To conclude we prove one more simple property of the vector π which explains the meaning of the term "stationary distribution". Assume that the initial distribution μ for the state ω_0 agrees with π. We show that $P(\omega_k = x^{(j)}) = \pi_j$ for any k, $0 \le k \le n$. For $k = 0$ this property follows from the definition. Let us assume that the property is proved for $k < k_0$. By the total probability formula we have

$$P(\omega_k = x^{(j)}) = \sum_{i=1}^{r} P(\omega_{k-1} = x^{(i)})P(\omega_k = x^{(j)}|\omega_{k-1} = x^{(i)})$$

$$= \sum_{i=1}^{r}\pi_i p_{ij} = \pi_j$$

by the equality $\pi P = \pi$. The statement we have proved means that if the initial distribution is π then the probability distribution for any ω_k is given by that very vector π and does not depend on k. Hence the word "stationary".

Markov chains first appeared in the work of the Russian mathematician A. A. Markov. Later on they appeared significantly in the interesting and deep theory of Markov processes. The main results here were obtained by A. N. Kolmogorov, A. Ia. Khinchin, Doeblin, K. L. Chung, and others.

5.5 Application to Products of Positive Matrices

Let $A = \|a_{ij}\|$ be a matrix with strictly positive entries, $1 \le i, j \le r$. Then $A^n = \|a_{ij}^{(n)}\|$ where

$$a_{ij}^{(n)} = \sum_{1 \le i_1, \ldots, i_{n-1} \le r} a_{ii_1} \cdot a_{i_1 i_2} \cdots a_{i_{n-2}i_{n-1}} \cdot a_{i_{n-1}j}.$$

We shall use the ergodic theorem for Markov chains in order to study the asymptotic behavior of $a_{ij}^{(n)}$ as $n \to \infty$.

Lemma 5.4. *There exist a positive number λ and column vectors $e = (e_1, \ldots, e_r)'$, $e^* = (e_1^*, \ldots, e_r^*)'$ such that*

1. $e_j > 0$, $e_j^* > 0$, $1 \leq j \leq r$;
2. $\sum_{j=1}^r a_{ij} e_j = \lambda e_i$, $1 \leq i \leq r$; and $\sum e_i^* a_{ij} = \lambda e_j^*$, $1 \leq j \leq r$.

Proof. Let us show that for any matrix A with positive entries there exist at least one vector e with positive coordinates, and $\lambda > 0$, for which

$$\sum_{j=1}^r a_{ij} e_j = \lambda e_i, \qquad 1 \leq i \leq r.$$

Consider a convex set $\mathcal{H}$ of vectors $h = (h_1, \ldots, h_r)$ such that $h_i \geq 0$, $1 \leq i \leq r$, and $\sum_{i=1}^r h_i = 1$. The matrix A determines a continuous transformation $\mathcal{A}$ of $\mathcal{H}$ into itself through the formula $\mathcal{A}h = h'$, where

$$h_i' = \frac{\sum_{j=1}^r a_{ij} h_j}{\sum_{i=1}^r \sum_{j=1}^r a_{ij} h_j}.$$

The Brouwer Theorem states that any such map has at least one fixed point. If e is this fixed point then $\mathcal{A}e = e$, i.e.

$$e_i = \frac{\sum_{j=1}^r a_{ij} e_j}{\sum_{i=1}^r \sum_{j=1}^r a_{ij} e_j}.$$

Letting $\lambda = \sum_{i=1}^r \sum_{j=1}^r a_{ij} e_j$ we get the desired result.

Consider the matrix $A^* = \|a_{ij}^*\|$, $a_{ij}^* = a_{ji}$. Then by the first part of Lemma 5.4 one can find λ^* and e^* such that $A^* e^* = \lambda^* e^*$, i.e.

$$\sum_{j=1}^r a_{ji} e_j^* = \lambda^* e_i^*,$$

and $e_i^* > 0$. The equalities

$$\lambda(e, e^*) = (Ae, e^*) = (e, A^* e^*) = \lambda^*(e, e^*)$$

show that $\lambda = \lambda^*$.   □

Using Lemma 5.4, put

$$p_{ij} = \frac{a_{ij} e_j}{\lambda e_i}.$$

It is easy to see that the matrix P is a stochastic matrix with strictly positive entries. The stationary distribution for this matrix is $\pi_i = e_i e_i^*$ provided that e and e^* are normalized in such a way that $\sum \pi_i = \sum e_i e_i^* = 1$. Indeed

$$\sum_i \pi_i p_{ij} = \sum_i e_i e_i^* \frac{a_{ij} e_j}{\lambda e_i} = \frac{1}{\lambda} e_j \sum e_i^* a_{ij} = e_j e_j^* = \pi_j.$$

We can rewrite $a_{ij}^{(n)}$ as follows:

$$a_{ij}^{(n)} = \sum a_{ii_1} a_{i_1 i_2} \cdots a_{i_{n-2} i_{n-1}} a_{i_{n-1} j}$$
$$= \lambda^n \sum p_{ii_1} p_{i_1 i_2} \cdots p_{i_{n-2} i_{n-1}} p_{i_{n-1} j} \cdot e_i \cdot e_j^{-1}$$
$$= \lambda^n e_i p_{ij}^{(n)} e_j^{-1}.$$

The ergodic theorem for stochastic matrices gives $p_{ij}^{(n)} \to \pi_j = e_j e_j^*$ as $n \to \infty$. Thus

$$\frac{a_{ij}^{(n)}}{\lambda^n} \to e_i \pi_j e_j^{-1} = e_i e_j^*$$

and the convergence in the last expression is exponentially fast. Thus

$$a_{ij}^{(n)} \sim \lambda^n e_i e_j^* \quad \text{as} \quad n \to \infty. \tag{5.1}$$

Now we can show that the vectors e, e^* and $\lambda > 0$ in Lemma 5.4 are unique. From the last expression,

$$\lim_{n \to \infty} \frac{1}{n} \ln a_{ij}^{(n)} = \ln \lambda.$$

Thus λ is uniquely determined by A.

Suppose that there exist two vectors $(e^*)'$ and $(e^*)''$ with the needed properties. Choose an arbitrary vector e from Lemma 5.4. Then the existence of the two vectors $(e^*)'$ and $(e^*)''$ implies that the matrix P has two stationary distributions, which is impossible. Thus $(e^*)' = (e^*)''$. Replacing A by A^* we get the same result for the vector e.

Relation (5.1), together with the statements about the unicity of e and e^*, give a complete description of the behavior of $a_{ij}^{(n)}$ as $n \to \infty$.

Lecture 6. Random Walks on the Lattice z^d

The topic of random walks is one of the most important and most developed of those studied in probability theory. Problems from a lot of applications of probability theory are connected to random walks. In fact we have already encountered one of them in the gambler's ruin problem. The theory of Markov chains can be viewed as a theory of random walks on graphs.

In this lecture we study random walks on the lattice $\mathbf{Z}^d$. The lattice $\mathbf{Z}^d$ is understood to be the collection of points $x = \{x_1, \ldots, x_d\}$ where $-\infty < x_i < \infty$ are integers and $1 \le i \le d$. A random walk on $\mathbf{Z}^d$ is a Markov chain whose state space is $X = \mathbf{Z}^d$. By the same token we have here an example of a Markov chain with a countable state space. Following the definitions of the previous lecture we construct the graph of the Markov chain - i.e. from each point $x \in \mathbf{Z}^d$ we find those $y \in \mathbf{Z}^d$ for which the probability p_{xy} of the transition $x \to y$ is positive. These transition probabilities must satisfy the relation $\sum_y p_{xy} = 1$. As in the case of finite Markov chains the probability distribution of the Markov chain is uniquely determined by the initial distribution $\mu = \{\mu_x, x \in \mathbf{Z}^d\}$ and the collection of transition probabilities $P = \|p_{xy}\|$ which now form an infinite stochastic matrix, sometimes called the stochastic operator of the random walk. Often the initial distribution will be taken to be the distribution which is concentrated at one point, i.e. $\delta^{(x_0)} = \{\delta_{x_0-x}, x \in \mathbf{Z}^d\}$ where $\delta_x = 1$ for $x = 0$, $\delta_x = 0$ for $x \ne 0$. This corresponds to considering trajectories of the random walk which begin at the point x_0.

Definition 6.1. A random walk is said to be spatially homogeneous if $p_{xy} = p_{y-x}$, where $p = \{p_z, z \in \mathbf{Z}^d\}$ is a probability distribution on the lattice $\mathbf{Z}^d$.

In the sequel, without mentioning it explicitly, we will only consider random walks which are spatially homogeneous. The theory of such random walks is closely connected to sequences of independent identical trials. Indeed let $\omega = \{\omega_0, \omega_1, \ldots, \omega_n\}$ be an elementary outcome for the random walk $p(\omega) = \mu_{\omega_0} p_{\omega_0 \omega_1} \cdots p_{\omega_{n-1} \omega_n} = \mu_{\omega_0} p_{\omega_1 - \omega_0} \cdots p_{\omega_n - \omega_{n-1}}$. We now find the probability distribution for the increments $\omega_1' = \omega_1 - \omega_0, \ldots, \omega_n' = \omega_n - \omega_{n-1}$. It can be obtained from the previous expression by a summation over ω_0; we simply get $p_{\omega_1 - \omega_0} \cdots p_{\omega_n - \omega_{n-1}}$. We arrive at the expression for the probability of a sequence of independent identical trials $p_{\omega_1'} p_{\omega_2'} \cdots p_{\omega_n'}$. We will make use of this property substantially.

One of the main concepts in the theory of random walks is that of recurrence. Let us take $\mu = \delta^{(0)}$, i.e. let us consider those random walks which begin at the point 0. Let $\omega = \{\omega_0, \omega_1, \ldots, \omega_k\}$ be a trajectory of the random walk, i.e. $\omega_0 = 0$, $p_{\omega_i - \omega_{i-1}} > 0$. We assume that $\omega_i \neq 0$, $1 \leq i \leq k$ and $\omega_k = 0$. In the case of such an ω we can say that the trajectory of the random walk returns to the initial point for the first time at the k-th step. The set of such ω will be denoted by $\Omega^{(k)}$. We will set, as usual, for $\omega \in \Omega^{(k)}$ $p(\omega) = p_{\omega_1 - \omega_0} \cdots p_{\omega_k - \omega_{k-1}}$, $f_k = \sum_{\omega \in \Omega^{(k)}} p(\omega)$, $k > 0$. It is also convenient to assume that $f_0 = 0$.

Definition 6.2. A random walk is said to be recurrent if $\sum_{k=1}^{\infty} f_k = 1$. If this latter sum is less than one the random walk is said to be transient.

The definition of recurrence means that the probability of those trajectories which return to the initial point is equal to 1. We introduce here a general criterion for the recurrence of a random walk. Let $\bar{\Omega}^{(k)}$ consist of those points $\omega = \{\omega_0, \omega_1, \ldots, \omega_k\}$ such that $\omega_0 = \omega_k = 0$. In this notation it is possible that $\omega_i = 0$ for some i, $1 \leq i \leq k$. Consequently $\Omega^{(k)} \subseteq \bar{\Omega}^{(k)}$. We set, as previously, $p(\omega) = \prod_{i=1}^{k} p_{\omega_i - \omega_{i-1}}$, for $\omega \in \bar{\Omega}^{(k)}$, and $p_k = \sum_{\omega \in \bar{\Omega}^{(k)}} p(\omega)$. It is also convenient to assume that $p_0 = 1$.

Lemma 6.1 (Criterion for Recurrence). *A random walk is recurrent if and only if $\sum_{k \geq 0} p_k = \infty$.*

Proof. We now introduce an important formula which relates $\{f_k\}$ and $\{p_k\}$:

$$p_k = f_k + f_{k-1} \cdot p_1 + \cdots + f_0 \cdot p_k. \tag{6.1}$$

We have $\bar{\Omega}^{(k)} = \bigcup_{i=1}^{k} C_i$, where $C_i = \{\omega | \omega \in \bar{\Omega}^{(k)}, \omega_i = 0$ and $\omega_j \neq 0, 1 \leq j \leq i\}$. Since the C_i are pairwise disjoint

$$p_k = \sum_{i=1}^{k} P(C_i).$$

We note furthermore that

$$P(C_i) = \sum_{\omega \in C_i} p_{\omega_1 - \omega_0} \cdots p_{\omega_i - \omega_{i-1}} p_{\omega_{i+1} - \omega_i} \cdots p_{\omega_k - \omega_{k-1}}$$

$$= \sum_{\omega \in \Omega^{(i)}} p_{\omega_1 - \omega_0} \cdots p_{\omega_i - \omega_{i-1}} \sum_{\omega \in \bar{\Omega}^{(k-i)}} p_{\omega_{i+1} - \omega_i} \cdots p_{\omega_k - \omega_{k-1}}$$

$$= f_i \cdot p_{k-i}.$$

By the same token, if we recall that $f_0 = 0$, $p_0 = 1$ we have

$$p_k = \sum_{i=0}^{k} f_i \cdot p_{k-i},$$

$$p_0 = 1. \tag{6.2}$$

This completes the proof of (6.1).

We now need to make a diversion and talk about generating functions. Let $\{a_n\}$ be an arbitrary bounded sequence, i.e. $|a_n| \leq$ const. The generating function of the sequence $\{a_n\}$ is the power series $A(z) = \sum_{n \geq 0} a_n z^n$ which is an analytic function of the complex variable z in the domain $|z| < 1$. An essential fact for us is that $A(z)$ uniquely determines the sequence $\{a_n\}$ since $a_n = \frac{1}{n!} A^{(n)}(0)$.

Returning to our random walk, we introduce the following generating functions:

$$F(z) = \sum_{k \geq 0} f_k z^k, \quad P(z) = \sum_{k \geq 0} p_k z^k.$$

Let us multiply the left and right sides of (6.2) by z^k and sum with respect to k from 0 to ∞. We then obtain on the left $P(z)$ and, on the right, as is easily seen, we obtain $1 + P(z) \cdot F(z)$, i.e.

$$P(z) = 1 + P(z) \cdot F(z).$$

So (6.1) and (6.2) imply that the generating functions are related by a very simple equation: $F(z) = 1 - 1/P(z)$. We now note that by Abel's Theorem

$$\sum_{k=1}^{\infty} f_k = F(1) = \lim_{z \to 1} F(z) = 1 - \lim_{z \to 1} \frac{1}{P(z)}.$$

In the latter equalities, and below, $z \to 1$ from the left on the real axis.

We first assume that $\sum_{k=1}^{\infty} p_k < \infty$. Then, by Abel's Theorem,

$$\lim_{z \to 1} P(z) = P(1) = \sum_{k=0}^{\infty} p_k < \infty$$

and $\lim_{z \to 1} (1/P(z)) = 1/\sum_{k=0}^{\infty} p_k > 0$. So $\sum_{k=0}^{\infty} f_k < 1$, i.e. the random walk is transient.

A slightly more complicated argument is used when $\sum_{k=0}^{\infty} p_k = \infty$. We show that in this case $\lim_{z \to 1} 1/P(z) = 0$. Let us fix $\epsilon > 0$. We find $N = N(\epsilon)$ such that $\sum_{k=0}^{N} p_k \geq 2/\epsilon$. Then for z sufficiently close to 1 we have $\sum_{k=0}^{N} p_k z^k \geq 1/\epsilon$. Consequently, for such z

$$\frac{1}{P(z)} \leq \frac{1}{\sum_{k=0}^{N} p_k z^k} \leq \epsilon.$$

This means that $\lim_{z \to 1} 1/P(z) = 1/\sum_{k=0}^{\infty} p_k = 0$. This completes the proof of the criterion. $\qquad \square$

The advantage of the criterion is that it reduces the question of whether a random walk is recurrent to a concrete problem connected with a sequence of independent identical trials. We now consider some applications of this criterion. Let us introduce the quantity

$$E = \sum_{z \in \mathbb{Z}^d} z p_z.$$

Clearly E is a $d-$dimensional vector equal to the mathematical expectation, i.e. the mean value of the one step displacements.

Theorem 6.1. Let $E \neq 0$ and $\sum_{z \in \mathbb{Z}^d} \|z\|^4 p_z < \infty$. Then the random walk is transient.

Proof. Intuitively the last statement should be clear, since it follows from the law of large numbers that ω_n/n should be close to E and that the probability that $\omega_n = 0$ should be small. We now support this fact by an accurate proof. The probability

$$p_n = \sum_{\omega': \sum_{k=1}^n \omega'_k = 0} p_{\omega'_1} p_{\omega'_2} \cdots p_{\omega'_n}.$$

Clearly

$$p_n \leq P\left\{ \left\| \sum_{k=1}^n \omega'_k - nE \right\| \geq \frac{nE}{2} \right\} = P\left\{ \left\| \sum_{k=1}^n \omega'_k - nE \right\|^4 \geq \frac{n^4 E^4}{16} \right\}.$$

Here the probability P is calculated by using the probability distribution which corresponds to the sequence of independent identical trials. By Chebyshev's inequality we have:

$$P\left\{ \left\| \sum_{k=1}^n \omega'_k - nE \right\|^4 \geq \frac{n^4 E^4}{16} \right\} \leq \frac{16. E \| \sum_{k=1}^n \omega'_k - nE \|^4}{E^4 n^4}.$$

We now show that $E\| \sum_{k=1}^n \omega'_k - nE\|^4 \leq \text{const} \cdot n^2$; this implies the inequality $p_n \leq \text{const}/n^2$, which gives transience. We have

$$\left\| \sum_{k=1}^n \omega'_k - nE \right\|^2 = \left\| \sum_{k=1}^n (\omega'_k - E) \right\|^2 = \sum_{k_1, k_2 = 1}^n (\omega'_{k_1} - E, \omega'_{k_2} - E)$$

and

$$\left\| \sum_{k=1}^n \omega'_k - nE \right\|^4 = \sum_{k_1, k_2, \ell_1, \ell_2 = 1}^n (\omega'_{k_1} - E, \omega'_{k_2} - E)(\omega'_{\ell_1} - E, \omega'_{\ell_2} - E).$$

We show that $E(\omega'_{k_1} - E, \omega'_{k_2} - E)(\omega'_{\ell_1} - E, \omega'_{\ell_2} - E)$ can be different from zero only in the case when no more than two indices among k_1, k_2, ℓ_1 and ℓ_2 are distinct. First we derive from this fact the inequality that we need. The number of such terms is no greater than $\text{const} \cdot n^2$. By the Cauchy-Bunyakovsky inequality we have

$$|E(\omega'_{k_1} - E, \omega'_{k_2} - E)(\omega'_{\ell_1} - E, \omega'_{\ell_2} - E)|$$
$$\leq E|(\omega'_{k_1} - E, \omega'_{k_2} - E)| \cdot |(\omega'_{\ell_1} - E, \omega'_{\ell_2} - E)|$$
$$\leq E\|\omega'_{k_1} - E\| \cdot \|\omega'_{k_2} - E\| \cdot \|\omega'_{\ell_1} - E\| \cdot \|\omega'_{\ell_2} - E\|$$
$$= E\|\omega'_k - E\|^2 \cdot \|\omega'_\ell - E\|^2,$$

where k and ℓ are the corresponding indices. If $k = \ell$ we have $E\|\omega'_k - E\|^4 = \sum_{z \in \mathbf{Z}^d} \|z - E\|^4 p_z < \infty$, which is easily verified from the conditions of the the-orem. If $k \neq \ell$ then, as we have seen when studying sequences of independent identical trials,

$$E\|\omega'_k - E\|^2 \cdot \|\omega'_\ell - E\|^2 = \sum p_{\omega'} \cdot \|\omega'_k - E\|^2 \cdot \|\omega'_\ell - E\|^2$$

$$= \sum_{\omega'_1, \dots, \omega'_n} p_{\omega'_1} \dots p_{\omega'_k} \dots p_{\omega'_\ell} \dots p_{\omega'_n} \|\omega'_k - E\|^2 \cdot \|\omega'_\ell - E\|^2$$

$$= \sum_{\omega'_k} p_{\omega'_k} \cdot \|\omega'_k - E\|^2 \cdot \sum_{\omega'_\ell} p_{\omega'_\ell} \cdot \|\omega'_\ell - E\|^2.$$

The finiteness of the last sum is also easily verified from the conditions of the theorem.

It remains to be checked that when at least three among the indices k_1, k_2, ℓ_1 and ℓ_2 are distinct the corresponding mathematical expectation is equal to 0. In this case at least one index appears only once. Assume that the index that appears only once is, for example, k_1. Then

$$E(\omega'_{k_1} - E, \omega'_{k_2} - E)(\omega'_{\ell_1} - E, \omega'_{\ell_2} - E)$$

$$= \sum p_{\omega'_1} \dots p_{\omega'_{k_1}} \dots p_{\omega'_{k_2}} \dots p_{\omega'_{\ell_1}} \dots p_{\omega'_{\ell_2}} \dots p_{\omega'_n} \cdot$$

$$\sum_{s=1}^{d} \left(\omega'_{k_1}(s) - E(s), \omega'_{k_2}(s) - E(s)\right)\left(\omega'_{\ell_1}(s) - E(s), \omega'_{\ell_2}(s) - E(s)\right).$$

Here we denote by $\omega'_k(s)$, $E(s)$, $1 \leq s \leq d$ the components of the vectors ω'_k and E respectively. Since the series converges absolutely we can perform the summation in the last sum in any order desired. Let us fix all ω'_i, $i \neq k$, and sum with respect to all values ω'_k. Since $\sum \omega'_k(s) p_{\omega'_k} = E(s)$ by definition of the vector E, the obtained sum is equal to 0. $\square$

Let $e_1, \dots, e_d$ be the unit coordinate vectors and let $p_{y-x} = 1/2d$ if $y = \pm e_s$, $1 \leq s \leq d$, and 0 otherwise. Such a random walk is said to be elementary.

Theorem 6.2 (Polya). *An elementary random walk is recurrent for $d = 1, 2$ and transient for $d \geq 3$.*

It is sometimes said that this theorem is the mathematical foundation for the saying "All roads lead to Rome". We note however that if we also include cosmic paths then this saying is not true.

The proof follows very simply from the criterion. The probability p_n is the probability that $\sum_{k=1}^{n} \omega'_k = 0$. For $d = 1$ this probability is investigated in the De Moivre-Laplace Local Limit Theorem; it behaves like $\text{const}/\sqrt{n}$. In the general case the probability decreases as $\text{const}/n^{d/2}$. For $d = 2$ or 3 these statements were proved in fact in Lecture 3. Indeed for $d = 2$ the increments ω'_k take four values, corresponding to $\pm e_s$, $s = 1, 2$, and the probability of each value is equal to $1/4$. Return to 0 occurs when the number of occurrences of e_1 and $-e_1$ are equal, as well as the number of occurrences of e_2 and $-e_2$. The probability of such an event was studied in Lecture 3, where it was shown (see Theorem 3.3) that $p_n \sim 4/2\pi n$ as $n \to \infty$. For $d = 3$ the increment ω_k' takes six values, $\pm e_s$, $s = 1, 2, 3$, and the probability of each of the values is equal to $1/6$. Return to 0 occurs when the number of occurrences of e_1 and $-e_1$ are equal, and similarly for e_2 and $-e_2$, and e_3 and $-e_3$. It follows from Theorem 3.3 that $p_n \leq \text{const}/n^{3/2}$ in this case, and this implies transience.

It follows easily from Polya's Theorem that in dimension $d = 3$ the "typical" trajectories of a random walk go off to infinity as $n \to \infty$. One can ask many questions about the asymptotic properties of such trajectories. For example, for each n consider the unit vector $\frac{\omega_n}{\|\omega_n\|} = \nu_n$; this consists in transposing the random walk to the unit sphere. One question is, for typical trajectories does $\lim_{n \to \infty} \nu_n$ exist? This would imply that a typical trajectory goes off to infinity in a given direction. It turns out that this is not the case, and there is no such limit. Furthermore the vectors ν_n are uniformly distributed on the unit sphere. This means that a typical trajectory, when going to infinity, is seen from the origin under a more or less arbitrary angle. Such a phenomenon is possible because of the fact that the trajectory of a random walk spreads out over a length of order $O(\sqrt{n})$ in n steps, and therefore manages to move in all directions.

Spatially homogeneous random walks on $\mathbf{Z}^d$ are special cases of homogeneous random walks on groups or sub-groups. Let G be a countable group and $p = \{p_g, g \in G\}$ be a probability distribution on the group G. We consider a Markov chain in which the space of states is the group G, and the transition probability is $p_{xy} = p_{yx^{-1}}$, $y \in G$, $x \in G$. As in the usual lattice $\mathbf{Z}^d$ we can formulate a definition for the recurrence of a random walk and prove an analogous criterion. The proof of this criterion makes use of limit theorems for random variables with values in a group G. In the case of elementary random walks the answer depends substantially on the group G. For example if G is taken to be the free group with two generators, a and b, and if the probability distribution p is concentrated on the four points a, b, a^{-1} and b^{-1}, then such a random walk will always be transient.

There are interesting problems in connection with continuous groups. The groups $SL(n, R)$ of matrices of order n with real elements, and determinant 1, arise particularly often. Special methods have been worked out to study random walks on such groups.

Lecture 7. Branching Processes

Here we consider another example of Markov chains, which arises in the analysis of so-called multiple decay processes for elementary particles. The theory of such processes was introduced in articles by the Soviet mathematicians A. N. Kolmogorov and B. A. Sevast'yanov.

In the simplest case we assume that we have a system of particles of one type, whose evolution is as follows: during a unit of time each particle either dies or remains unchanged or splits into several particles of the same type. We also assume that at time 0 one particle is present and that the whole evolution is considered on the time interval $[0, n]$. As always we begin with the construction of the space $\Omega^{(n)}$ of elementary outcomes. An individual outcome $\omega^{(n)} \in \Omega^{(n)}$ describes all consecutive transformations of the particles, and it is convenient to represent it in the form of a genealogical tree. The next figure represents an example of an $\omega^{(3)}$ for $n = 3$.

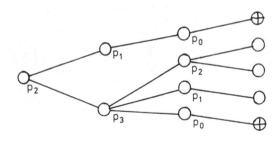

Fig. 7.1.

A circle with a cross means that the particle has died (disappeared). Since each particle can split into any number of particles, $\Omega^{(n)}$ in general is countable. Genealogical trees are possible which end with one cross. In this case the whole process stops. We then consider that the genealogical tree is extended by extending each cross into another cross. The number n is called the number of generations. It therefore makes sense to talk of particles of the k-th generation, $0 \leq k \leq n$.

Let $\{p_k\}_0^\infty$ be a probability distribution where each probability p_k gives the probability of a particle splitting into k particles ($k = 0$ corresponds to death, $k = 1$ corresponds to the intact preservation of the particle). The probability distribution on $\Omega^{(n)}$ is constructed from the fact that the individual particles split independently. Formally, this means the following: next to each particle x of the $m-$th generation (a cross does not represent a particle), $0 \leq m \leq n-1$, we write $p(x) = p_k$ if the particle splits into k particles, $k = 0, 1, 2, \ldots$. Next to each cross we write 1. We now set

$$p(\omega^{(n)}) = \prod_{x \in \omega^{(n)}} p(x).$$

The fact that in the latter expression the product over all particles x up to the $(n-1)-$st generation (inclusive) appears, reflects the independence of the splitting process for all particles.

Lemma 7.1. *The numbers $p(\omega^{(n)})$ define a probability distribution on $\Omega^{(n)}$.*

Proof. The proof is carried out by induction on n. For $n = 1$ the statement follows from the fact that the set $\{p_k\}_0^\infty$ is a probability distribution. Let us assume that the statement is already proved for a given n. Then

$$\sum_{\omega^{(n+1)} \in \Omega^{(n+1)}} p(\omega^{(n+1)}) = \sum_{\omega^{(n)} \in \Omega^{(n)}} \sum_{\omega^{(n+1)} \to \omega^{(n)}} p(\omega^{(n+1)}),$$

where the notation $\omega^{(n+1)} \to \omega^{(n)}$ means that $\omega^{(n+1)} \in \Omega^{(n+1)}$ was obtained from $\omega^{(n)} \in \Omega^{(n)}$. We note now that $p(\omega^{(n+1)}) = p(\omega^{(n)}) \prod p(x)$, where the last product is carried out over all particles of the $n-$th generation in $\omega^{(n)}$. If x is a cross then $p(x) = 1$ by our assumption. If x is a particle then it can split into any number of particles. Therefore

$$\sum_{\omega^{(n+1)} \to \omega^{(n)}} p(\omega^{(n+1)}) = p(\omega^{(n)}) \sum_{\omega^{(n+1)} \to \omega^{(n)}} \prod_x p(x)$$

$$= p(\omega^{(n)}) \prod_x \sum_{k=0}^\infty p_k = p(\omega^{(n)}).$$

By the same token

$$\sum_{\omega^{(n+1)} \in \Omega^{(n+1)}} p(\omega^{(n+1)}) = \sum_{\omega^{(n)} \in \Omega^{(n)}} p(\omega^{(n)}) = 1$$

by the induction hypothesis. $\square$

The space $\Omega^{(n)}$ with the probability distribution $\{p^{(n)}\}$ is called the branching process. Let us denote by q_n the probability of those $\omega^{(n)}$ which end only with crosses - i.e. corresponding to the extinction of our process. It is clear that $q_{n+1} \geq q_n$.

Definition 7.1. The branching process is said to be degenerate if $\lim_{n \to \infty} q_n = 1$. In the opposite case the process is said to be non-degenerate.

We now obtain a simple criterion for the degeneracy of the process. Let us introduce the quantity

$$m = \sum_{k=0}^{\infty} k p_k,$$

which represents the mean number of progeny of one particle. We have the following theorem.

Theorem 7.1. *Let $p_1 < 1$. The branching process is degenerate if and only if $m \leq 1$.*

Remark. The statement of the theorem for $m < 1$ appears quite natural. The fact that it holds for $m = 1$ is more surprising.

Proof. As in the previous lecture we use generating functions. Let us introduce the random variable $\nu^{(n)}(\omega^{(n)})$ equal to the number of particles of the n-th generation in $\omega^{(n)}$. It is clear that $\nu^{(n)}(\omega^{(n)})$ takes the values 0 (extinction), $1, 2, \ldots$. Let us form the generating function $\phi^{(n)}(z) = \sum_{s=0}^{\infty} P\{\nu^{(n)}(\omega^{(n)}) = s\} z^s$. Then $q_n = \phi^{(n)}(0)$. We also set $\phi(z) = \sum_{k=0}^{\infty} p_k z^k$. We study in more detail the properties of the function $\phi(z)$ for $0 \leq z \leq 1$. Since $\sum_{k=0}^{\infty} p_k = 1$, $1 = \lim_{z \to 1} \phi(z)$. Moreover $\phi'(z) = \sum_{k=1}^{\infty} k p_k z^{k-1} \geq 0$, where the value of 0 in the last expression can occur for $0 < z \leq 1$ only in the degenerate case where $p_0 = 1$, that is where no splitting occurs and the particle dies immediately. Excluding this case from consideration we find that $\phi'(z) > 0$ for $0 < z \leq 1$, i.e. $\phi(z)$ is strictly monotonic. Furthermore $\phi''(z) = \sum_{k=2}^{\infty} k(k-1) p_k z^{k-2} \geq 0$, and for $0 < z \leq 1$ equality is possible only in the case where $p_0 + p_1 = 1$ and $p_k = 0$ for $k > 1$, i.e. when $\phi(z) = p_0 + p_1 z$ is a linear function. For $p_1 = 1$, $p_0 = 0$ its graph is the bisectrix $\phi(z) = z$, and for $p_1 < 1$ its graph intersects the bisectrix at one point. For $p_0 + p_1 < 1$, $\phi''(z) > 0$ for $0 < z \leq 1$ and therefore $\phi(z)$ is a strictly convex function. Since $\phi(1) = 1$, $\phi(0) \leq 1$, the graph of $\phi(z)$ intersects the bisectrix $\phi(z) = z$ at one point at most for which $0 < z < 1$.

Lemma 7.2. *If $m \leq 1$ the graph of $\phi(z)$ intersects the bisectrix $\phi = z$ for $0 \leq z \leq 1$ only at the point $z = 1$. If $m > 1$ there exists a z_0, $0 < z_0 < 1$, such that $\phi(z_0) = z_0$.*

Remark. It follows from our previous analysis that the point z_0 is unique.

Remark. The statement of the lemma is obvious for $p_k = 0$, $k \geq 2$, i.e. $\phi(z) = p_0 + p_1 z$.

Proof. We first show that for $m \leq 1$ the point z_0 does not exist. Assume that this is not the case, and that for some $z_0 < 1$ we have $\phi(z_0) = z_0$. By Rolle's Theorem there exists ξ such that $\phi(1) - \phi(z_0) = 1 - z_0 = \phi'(\xi)(1 - z_0)$, i.e. $\phi'(\xi) = 1$. But the function $\phi''(z) > 0$ for $0 < z \leq 1$ (we do not consider the degenerate case here). Therefore $\phi'(z)$ is strictly monotonic. Consequently we must have $\lim_{z \to 1} \phi'(z) > 1$, which is impossible, since by Abel's Theorem $\lim_{z \to 1} \phi'(z) = \sum_{k=1}^{\infty} k p_k \leq 1$.

There remains the case when $m > 1$. Consider $\phi_1(z) = \phi(z) - z$. We have $\phi_1(0) = \phi(0) = p_0 \geq 0$, $\phi_1(1) = \phi(1) - 1 = 0$. In the trivial case where $p_0 = 0$ extinction and, by the same token, degeneracy are impossible. We will therefore assume that $\phi_1(0) = p_0 > 0$. Since $\phi_1'(z) > 0$ in a left-half neighborhood of $z = 1$, the function $\phi_1(z)$ is negative in a left neighborhood of $z = 1$. Consequently there exists a z_0 such that $\phi_1(z_0) = 0$, i.e. $\phi(z_0) = z_0$.

$\square$

The proof of the theorem now consists in showing that $\lim_{n \to \infty} q_n = z_0$. We first introduce the following relations, which connect $\phi^{(n)}(z)$ and $\phi^{(n+1)}(z)$.

Lemma 7.3. $\phi^{(n+1)}(z) = \phi^{(n)}(\phi(z))$.

Proof. We have

$$\phi^{(n+1)}(z) = \sum_{s=0}^{\infty} P\{\nu^{(n+1)}(\omega^{(n+1)}) = s\} z^s$$

$$= \sum_{\omega^{(n)} \in \Omega^{(n)}} \sum_{\omega^{(n+1)} \to \omega^{(n)}} z^s p(\omega^{(n+1)})$$

$$= \sum_{\omega^{(n)} \in \Omega^{(n)}} p(\omega^{(n)}) \sum_{\omega^{(n+1)} \to \omega^{(n)}} \prod_x p(x) z^{\xi(x)}.$$

Here we use the same relation between $p(\omega^{(n)})$ and $p(\omega^{(n+1)})$ as before; the product $\prod_x p(x) z^{\xi(x)}$ is carried out over all particles of the n-th generation, and $\xi(x)$ is equal to ℓ only in the case where x has split into ℓ particles. Therefore $\nu^{(n+1)}(\omega^{(n+1)}) = \sum_x \xi(x)$, where the sum is carried out similarly, over the particles of the n-th generation. In particular, if x is a cross, then $p(x) = 1$ and $\xi(x) = 0$. Furthermore

$$\sum_{\omega^{(n+1)} \to \omega^{(n)}} \prod_x p(x) z^{\xi(x)} = \prod_x \sum_k p_k z^k = (\phi(z))^{\nu^{(n)}(\omega^{(n)})},$$

since $\nu^{(n)}(\omega^{(n)})$ is equal exactly to the number of particles of the n-th generation. By the same token

$$\phi^{(n+1)}(z) = \sum_{\omega^{(n)} \in \Omega^{(n)}} p(\omega^{(n)}) \cdot (\phi(z))^{\nu^{(n)}(\omega^{(n)})}$$

$$= \sum_{s=0}^{\infty} \sum_{\omega^{(n)} | \nu^{(n)}(\omega^{(n)}) = s} p(\omega^{(n)}) \cdot (\phi(z))^{\nu^{(n)}(\omega^{(n)})}$$

$$= \sum_{s=0}^{\infty} (\phi(z))^s P\{\nu^{(n)}(\omega^{(n)}) = s\} = \phi^{(n)}(\phi(z)).$$

□

For further discussion another relation will be of use, which follows from Lemma 7.3: $\phi^{(n+1)}(z) = \phi(\phi^{(n)}(z))$. For $n = 2$ both relations agree since $\phi^{(1)}(z) = \phi(z)$. We again argue by induction over n. It follows from Lemma 7.3 and from the induction hypothesis, and again from Lemma 7.3, that

$$\phi(\phi^{(n)}(z)) = \phi(\phi^{(n-1)}(\phi(z))) = \phi^{(n)}(\phi(z)) = \phi^{(n+1)}(z).$$

We now finish the proof of the theorem. We have already seen that $q_n = \phi^{(n)}(0) \leq q_{n+1} = \phi^{(n+1)}(0) \leq 1$. Therefore $\lim_{n \to \infty} q_n = \lim_{n \to \infty} \phi^{(n)}(0) = z_0 \leq 1$ exists. Since $\phi^{(n+1)}(0) = \phi(\phi^{(n)}(0))$, then

$$\phi(z_0) = \lim_{n \to \infty} \phi(\phi^{(n)}(0)) = \lim_{n \to \infty} \phi^{(n+1)}(0) = z_0,$$

i.e. $z_0 = \phi(z_0)$. We now show that z_0 is the smallest root of the equation $z_0 = \phi(z_0)$. Indeed let $\bar{z}_0$ be the smallest root. We do not consider the trivial case where $\bar{z}_0 = 0$, i.e. $p_0 = 0$. By the strong monotonicity of $\phi(z)$ we have

$$\phi(0) = p_0 < \phi(\bar{z}_0) = \bar{z}_0,$$

$$\phi^{(2)}(0) = \phi(\phi(0)) < \phi(\phi(\bar{z}_0)) = \phi(\bar{z}_0) = \bar{z}_0,$$

and by induction on n

$$\phi^{(n)}(0) = \phi(\phi^{(n-1)}(0)) < \phi(\bar{z}_0) = \bar{z}_0,$$

i.e. all $\phi^{(n)}(0) = q_n < \bar{z}_0$. But then $\lim_{n \to \infty} \phi^{(n)}(0)$ can be only $\bar{z}_0$.

We have already seen (Lemma 7.2) that for $m \leq 1$ the smallest root of the equation $\phi(z_0) = z_0$ is $z_0 = 1$, i.e. the process is degenerate, and for $m > 1$ there exists a unique root for the equation $\phi(z_0) = z_0$, $z_0 < 1$.   □

Problem. Let $m < 1$. Then according to Theorem 7.1 the branching process is degenerate. Let us construct a new space of elementary outcomes Ω whose points ω are the degenerate genealogical trees. For each ω the probability $p(\omega)$ is defined as before. We denote by $\tau(\omega)$ the number of generations in the tree ω. Show that $E\tau(\omega) < \infty$.

Lecture 8. Conditional Probabilities and Expectations

Let $(\Omega, \mathcal{F}, P)$ be a probability space, $C \in \mathcal{F}$ and $P(C) > 0$. The conditional distribution on the set C is a probability distribution defined on the σ−algebra $\mathcal{F}$ by

$$P(A|C) = \frac{P(A \cap C)}{P(C)}, \quad A \in \mathcal{F}.$$

This conditional distribution is in fact concentrated on the subsets of C. Given any random variable ξ, its mathematical expectation, computed with respect to this probability distribution, is the conditional expectation $E(\xi|C)$ given C. If $\xi = \{C_1, \ldots, C_r\}$ is an arbitrary finite partition of Ω we have the following objects:

a_1) the conditional distribution, conditional on each C_i, $1 \le i \le r$;
a_2) the space of elements of the partition $\Omega|\xi$ which consists of r points (the r elements of the partition) and a probability distribution on this space for which each probability is equal to $P(C_i)$.

These two objects are connected by the total probability formula

$$P(A) = \sum_{i=1}^{r} P(A|C_i)P(C_i), \tag{8.1}$$

which can be interpreted in the following way: first, the conditional measure of A is calculated with respect to each element C_i of our partition. The result is a function of i, i.e. a function on the space $\Omega|\xi$ of elements of the partition ξ, and the sum is the integral with respect to that space. Viewed in that way the total probability formula reduces the calculation of a probability to a procedure such as the one used when writing a double integral as an iteration of two integrals. The advantage is that one can often obtain some information or other about the individual terms. The formula

$$E\eta = \sum_{i=1}^{r} E(\eta|C_i)P(C_i), \tag{8.1'}$$

is called the total expectation formula and has exactly the same meaning.

It is now already clear in which direction one should generalize (8.1) and (8.1'). Let ξ be an arbitrary, not necessarily finite, partition of Ω into non-intersecting sets. The elements of the partition ξ will be denoted by C_ξ and the element of the partition which contains x will be denoted by $C_\xi(x)$. Each partition ξ defines a σ-algebra $\mathcal{F}(\xi)$. A set $C \in \mathcal{F}(\xi)$ if and only if there exists a union $C' \in \mathcal{F}$ of elements of the partition ξ and such that $P(C \Delta C') = 0$ (in probability theory, as well as in general measure theory, events which differ by a subset of measure 0 are identified). The space whose points are the elements C_ξ of the partition ξ is called the quotient space $\Omega|\xi$. Since $\mathcal{F}(\xi) \subseteq \mathcal{F}$ the probability distribution P induces a probability distribution on $\Omega|\xi$. So to each partition ξ corresponds a new probability space $(\Omega|\xi, \mathcal{F}(\xi), P)$. This is a natural generalization of a_2).

Important Example of a Partition. Let $\{\eta_n\}$ be a sequence of random variables, $n = 0, 1, 2, \ldots$. Let us take two numbers $n_1, n_2, (n_1 \leq n_2)$ and let us fix the values of of $\eta_{n_1}, n_1 \leq n \leq n_2$. We can obtain a partition $\xi_{n_1}^{n_2}$ in the following way: two points ω', ω'' belong to the same element of the partition $\xi_{n_2}^{n_1}$ if and only if $\eta_n(\omega') = \eta_n(\omega'')$ for all $n_1 \leq n \leq n_2$. One sometimes says that the partition $\xi_{n_1}^{n_2}$ is connected with the behavior of the sequence η_n on the time interval $[n_1, n_2]$. We can also consider the partition $\xi_{n_1}^\infty$ corresponding to the behavior of sequence η_n for $n \geq n_1$.

It is clear that we could consider finite collections of random variables $\eta_1, \ldots \eta_n$ and obtain partitions by fixing the values of some of the random variables, for example by fixing the values of $\eta_{n_1+1}, \ldots \eta_n$ where $n = n_1 + n_2$, that is, by fixing the values of the last n_2 random variables.

Now assume that we can introduce, on each C_ξ, a probability distribution $P(\cdot|C_\xi)$ defined on the σ-algebra $\mathcal{F}$ such that the total probability formula holds: for each $A \in \mathcal{F}$

$$P(A) = \int_{\Omega|\xi} P(A|C_\xi) dP(C_\xi), \tag{8.2}$$

or that the total expectation formula holds: for any random variable η

$$E\eta = \int_{\Omega|\xi} E(\eta|C_\xi) dP(C_\xi). \tag{8.2'}$$

We make the following remarks about (8.2) and (8.2'):

1) for a fixed A the quantity $P(A|C_\xi)$ or, for a fixed η, the quantity $E(\eta|C_\xi)$, is a function on the quotient space $\Omega|\xi$;
2) the notation $dP(C_\xi)$ refers to the induced probability distribution on $\Omega|\xi$;
3) the integral in (8.2) and (8.2') is the general Lebesgue integral with respect to the probability space $(\Omega|\xi, \mathcal{F}(\xi), P)$.

We also assume that $P(A|C_\xi)$ and $E(\eta|C_\xi)$ are measurable functions in that space.

Definition 8.1. The partition ξ is said to be measurable if the probability distribution $P(\cdot|C_\xi)$, called the conditional distribution, is defined, except for a subset of elements of the partition of measure 0, and the formulae (8.2) and (8.2') hold.

The probability $P(A|C_\xi)$ is called the conditional probability of the event A given C_ξ and $E(\eta|C_\xi)$ is called the conditional expectation. Formula (8.2′) is called the total expectation formula.

One can show that if ξ is a measurable partition the system of conditional probabilities is essentially (i.e. up to subsets of measure 0; we do not go into further details here) uniquely defined.

The partitions encountered in most problems of probability theory are measurable. At the same time it is easy to find an example of a non-measurable partition.

Example of a Non-measurable Partition. Let $\Omega = S^1$ be the unit circle and $\mathcal{F}$ be the σ−algebra of Borel subsets of the circle, and P be the Lebesgue measure on S^1. Let α be a fixed irrational number. Two points $\omega', \omega'' \in S^1$ belong to the same element of the partition ξ if and only if for some integers k_1, k_2, $\omega' - \omega'' = k_1 + k_2\alpha$.

We will not prove here that this partition is non-measurable. We simply note that non-measurable partitions are encountered in many important problems in the theory of non-commutative algebras and that the problem of their description and classification is of considerable current interest, and is the subject of intensive investigations nowadays.

We now look at the problem of constructing conditional distributions in one simple, but important, particular case. Let η_1 and η_2 be two random variables with joint density $p(x,y)$. This means that $p(x,y) \geq 0$, $\int \int p(x,y)\, dx\, dy = 1$ and

$$P\{\eta_1(\omega) \in A, \eta_2(\omega) \in B\} = \int_A \int_B p(x,y)\, dx\, dy,$$

where A, B are Borel subsets of the real line. Taking $B = \mathbb{R}^1$ we obtain

$$P\{\eta_1(\omega) \in A\} = \int_A \int_{-\infty}^\infty p(x,y)\, dx\, dy = \int_A p_1(x)\, dx,$$

where $p_1(x) = \int_{-\infty}^\infty p(x,y)\, dy$, i.e. η_1 has a distribution with density $p_1(x)$ (from now on we assume that all interchanges of the order of integration are valid). In just the same way η_2 has a distribution with density

$$p_2(y) = \int_{-\infty}^\infty p(x,y)\, dx.$$

We now find the conditional distribution of η_1, conditional on η_2. This means that we consider the partition ξ_2 obtained by fixing the value $\eta_2 = y$. The elements of C_{ξ_2} are parameterized by the value y. The quotient space $\Omega|\xi_2$ can be easily constructed. Any set C in $\mathcal{F}(\xi_2)$ is a union of elements C_{ξ_2}, i.e. is obtained by finding the set of those values y for which $C_{\xi_2} \subset C$. So $\mathcal{F}(\xi_2)$ is in fact the σ−algebra of Borel subsets of the line and of subsets of the line which differ from them by sets of measure 0. The induced probability distribution on the line is the probability distribution with density p_2. We set

$q(x|y) = p(x,y)/p_2(y)$ for those x,y such that $p_2(y) > 0$ and 0 for other x,y. Then $q(x|y) \geq 0$ and $\int q(x|y)\,dx = 1$ for such y. In other words, $q(x|y)$ is a probability density. For any Borel subset $A \subset R^1$

$$P\{\xi_1 \in A\} = \int_A \int_{-\infty}^{\infty} p(x,y)\,dx\,dy = \int_{-\infty}^{\infty} p_2(y)\,dy \int_A q(x|y)\,dx.$$

This is formula (8.2). It shows that η_1, for a fixed value y of the random variable η_2, has density $q(x|y)$. By unicity of the conditional distribution we now have found the conditional distribution of η_1 for a fixed η_2. We represent $p(x,y)$ in the form

$$p(x,y) = p_2(y) \cdot q(x|y), \tag{8.3}$$

where $p_2(y)$ is a probability density and $q(x|y)$ is a probability density in x for every y such that $p_2(y) > 0$. If such a representation is obtained for a given distribution then, by the same token, we have constructed the system of conditional probabilities.

Relation (8.3) can be directly generalized to the case of several η_i. Namely, let $n = n_1 + n_2$ and let $\eta_1, \ldots, \eta_n$ be n random variables whose joint distribution is given by the density $p(t_1, \ldots, t_n)$. Assume that p is represented in the form

$$
\begin{aligned}
p(x_1, \ldots, x_n, y_1, \ldots, y_n) = &\, p_2(y_1, \ldots, y_{n_2}) \cdot \\
&\, q(x_1, \ldots, x_{n_1} | y_1, \ldots, y_{n_2}),
\end{aligned}
\tag{8.4}
$$

where p_2 is a probability density and $q(x_1, \ldots, x_{n_1} | y_1, \ldots, y_{n_2})$ is a probability density in the variables $x_1, \ldots, x_{n_1}$ for any $y_1, \ldots, y_{n_2}$, such that $p_2(y_1, \ldots, y_{n_2}) > 0$. Then p_2 is the joint density of the random variables $\eta_{n_1+1}, \ldots, \eta_n$, and $q(x_1, \ldots, x_{n_1} | y_1, \ldots, y_{n_2})$ is the joint density of the random variables $\eta_1, \ldots, \eta_{n_1}$ for fixed values $y_1, \ldots, y_{n_2}$ of the random variables $\eta_{n_1+1}, \ldots, \eta_n$.

We now proceed to the most general approach to the construction of conditional probabilities and conditional expectations. Let $\mathcal{F}' \subset \mathcal{F}$ be an arbitrary σ−subalgebra of the σ−algebra $\mathcal{F}$. The question of whether $\mathcal{F}'$ can be represented as $\mathcal{F}' = \mathcal{F}(\xi)$ for a measurable partition ξ is far from trivial. It can be answered in the affirmative under some further assumptions on the probability space $(\Omega, \mathcal{F}, P)$ such as separability and completeness (in so-called Lebesgue spaces), but in general the answer may be negative in more complicated cases. Let $\eta \geq 0$ be a non-negative random variable such that $\int \eta(\omega)\,dP < \infty$.

Definition 8.2. A conditional expectation of the random variable η with respect to the σ−subalgebra $\mathcal{F}'$ is a function $E(\eta|\mathcal{F}')$ which is measurable with respect to the σ−algebra $\mathcal{F}'$ such that for any $A \in \mathcal{F}'$

$$\int_A \eta(\omega)\,dP = \int_A E(\eta|\mathcal{F}')\,dP.$$

The subtlety in the last equation lies in the fact that on the left hand side we have a probability distribution given on the whole σ−algebra $\mathcal{F}$, and on the right hand side a probability distribution given only on the $\sigma-$algebra $\mathcal{F}'$, i.e. a totally different object.

The existence of the conditional expectation is established by using the Radon-Nycodym Theorem. Specifically, $\int_A \eta(\omega)\, dP$ is a $\sigma-$additive set function defined on $\mathcal{F}'$. It is absolutely continuous with respect to P, since it follows from $P(A) = 0$ that $\int_A \eta(\omega)\, dP(\omega) = 0$. Then by the Radon-Nycodym Theorem there exists a function $g(\omega)$ which is measurable with respect to $\mathcal{F}'$ such that

$$\int_A \eta(\omega)\, dP = \int_A g(\omega)\, dP, \quad A \in \mathcal{F}'.$$

This function is taken to be the conditional expectation. If $\mathcal{F}' = \mathcal{F}(\xi)$ for a measurable partition ξ then the measurability of g with respect to $\mathcal{F}'$ means that g is a function on the quotient space $\Omega|\xi$, i.e. is constant on the elements of the partition ξ. This is the meaning of conditional expectation.

If we take for η the indicator of the set C, i.e. the random variable $\chi_C(\omega) = 1$ for $\omega \in C$ and $\chi_C(\omega) = 0$ for $\omega \notin C$, we obtain the definition of the conditional probability of the event C with respect to the $\sigma-$subalgebra $\mathcal{F}'$, denoted by $P(C|\mathcal{F}')$.

Given an arbitrary, not necessarily non-negative, random variable η, we may write it in the form $\eta = \eta_1 - \eta_2$, where $\eta_1 \geq 0$, $\eta_2 \geq 0$. We assume that $E\eta_1 < \infty$, $E\eta_2 < \infty$. The conditional expectation is then $E(\eta|\mathcal{F}') = E(\eta_1|\mathcal{F}') - E(\eta_2|\mathcal{F}')$.

Unfortunately, in the general case it is not possible to construct conditional distributions as a compatible system of $\sigma-$additive probability distributions. The further restrictions mentioned above are necessary for this.

Lecture 9. Multivariate Normal Distributions

In this lecture we consider a common example of a multivariate distribution. We begin with the non-degenerate case. Let $\xi_1, \ldots, \xi_n$ be n random variables whose joint distribution is given by the density $p(x_1, \ldots, x_n)$, of the form

$$p(x) = p(x_1, \ldots, x_n) = C e^{-\frac{1}{2}(A(x-m),(x-m))}. \tag{9.1}$$

Here C is a normalizing constant, $x = (x_1, \ldots, x_n)$ is an n-dimensional vector, $m = (m_1, \ldots, m_n)$ is also an n-dimensional vector, and A is a symmetric matrix. The density (9.1) is said to be the density of a non-degenerate multivariate normal distribution. The convenience of (9.1) lies in the fact that the density is defined in terms of very simple parameters: an n-dimensional vector m and a symmetric matrix A of order n.

Since $p(x)$ is integrable, $p(x) \to 0$ as $x \to \infty$. This is possible only in the case when A is a positive definite matrix. We now find the value of the constant C. By the normalization condition we have

$$\frac{1}{C} = \int_{-\infty}^{\infty} \cdots \int_{-\infty}^{\infty} e^{-\frac{1}{2}(A(x-m),(x-m))} dx_1 \ldots dx_n.$$

We first make the change of variables $y = x - m$:

$$\frac{1}{C} = \int_{-\infty}^{\infty} \cdots \int_{-\infty}^{\infty} e^{-\frac{1}{2}(Ay,y)} dy.$$

We now find an orthogonal matrix S such that $S^*AS = D$, where D is a diagonal matrix with elements d_i on the diagonal. We also have $\det D = \det A$. We now make the linear change of variables $y = Sz$. Then $(Ay,y) = (ASz, Sz) = (S^*ASz, z) = (Dz, z)$ (for an orthogonal matrix $S^* = S^{-1}$) and

$$\frac{1}{C} = \int_{-\infty}^{\infty} \cdots \int_{-\infty}^{\infty} e^{-\frac{1}{2}(Dz,z)} dz = \int_{-\infty}^{\infty} \cdots \int_{-\infty}^{\infty} e^{-\frac{1}{2}\sum_{i=1}^{n} d_i z_i^2} dz_1 \ldots dz_n$$

$$= \prod_i \int_{-\infty}^{\infty} e^{-\frac{1}{2} d_i z_i^2} dz_i = \prod_i \sqrt{2\pi d_i^{-1}} = (2\pi)^{n/2} \left(\prod_i d_i \right)^{-1/2}$$

$$= (2\pi)^{n/2} (\det D)^{-1/2} = (2\pi)^{n/2} (\det A)^{-1/2}.$$

Therefore $C = (2\pi)^{-n/2} \sqrt{\det A}$.

We now elucidate the probabilistic meaning of the vector m and the matrix A. Let us find the expectation $E\xi_i$:

$$E\xi_i = \int_{-\infty}^{\infty} \ldots \int_{-\infty}^{\infty} x_i\, p(x_1,\ldots,x_n)\, dx_1 \ldots dx_n$$

$$= C \int_{-\infty}^{\infty} \ldots \int_{-\infty}^{\infty} x_i\, e^{-\frac{1}{2}(A(x-m),(x-m))}\, dx_1 \ldots dx_n$$

$$= C \int_{-\infty}^{\infty} \ldots \int_{-\infty}^{\infty} (x_i - m_i)\, e^{-\frac{1}{2}(A(x-m),(x-m))}\, dx_1 \ldots dx_n + m_i$$

$$= C \int_{-\infty}^{\infty} \ldots \int_{-\infty}^{\infty} y_i\, e^{-\frac{1}{2}(Ay,y)}\, dy + m_i.$$

Here we made the change of variable $x - m = y$. The resulting integral is equal to 0, since the integrand is an odd function of y_i. Thus $m_i = E\xi_i$, i.e. the components of the vector m are the expectations of the random variables ξ_i.

We now find $\mathrm{Cov}(\xi_i, \xi_j)$. We have

$$\mathrm{Cov}(\xi_i, \xi_j) = E(\xi_i - m_i)(\xi_j - m_j)$$

$$= C \int_{-\infty}^{\infty} \ldots \int_{-\infty}^{\infty} (x_i - m_i)(x_j - m_j)\, e^{-\frac{1}{2}(A(x-m),(x-m))}\, dx$$

$$= C \int_{-\infty}^{\infty} \ldots \int_{-\infty}^{\infty} y_i\, y_j\, e^{-\frac{1}{2}(Ay,y)}\, dy_1 \ldots dy_n.$$

As before, we make the change of variable $y = Sz$ or, more explicitly, $y_i = \sum_{k=1}^{n} s_{ik} z_k$. Then

$$\mathrm{Cov}(\xi_i, \xi_j) = C \int_{-\infty}^{\infty} \ldots \int_{-\infty}^{\infty} \sum_{k,\ell} s_{ik} s_{j\ell} z_k z_\ell\, e^{-\frac{1}{2}\sum_i d_i z_i^2}\, dz_1 \ldots dz_n$$

$$= C \int_{-\infty}^{\infty} \ldots \int_{-\infty}^{\infty} \sum_{k} s_{ik} s_{jk} z_k^2\, e^{-\frac{1}{2}\sum_i d_i z_i^2}\, dz_1 \ldots dz_n$$

$$= \sum_{k=1}^{n} s_{ik} s_{jk} d_k^{-1}.$$

Here we used the fact that $\int_{-\infty}^{\infty} z_k z_\ell\, e^{-1/2 \sum_i d_i z_i^2}\, dz_1 \ldots dz_n = 0$ for $k \neq \ell$ and equals $\sqrt{2\pi} d_k^{-3/2}$ for $k = \ell$. We now note that $\sum_k s_{ik} d_k^{-1} s_{jk}$ is an element of the matrix $SD^{-1}S^* = SD^{-1}S^{-1} = A^{-1}$. Thus the matrix A^{-1} has an immediate probabilistic meaning. Its elements are the covariances of the random variables ξ_i, ξ_j. Let us assume that the covariances of the random variables ξ_i, ξ_j are equal to 0 for $i \neq j$. This means that A^{-1} is a diagonal matrix. Then A is also a diagonal matrix, and $p(x_1,\ldots,x_n) = p_1(x_1) \ldots p_n(x_n)$ is the product of one-dimensional normal densities, i.e. the random variables $\xi_1,\ldots,\xi_n$ are independent. Thus in the case of the multivariate normal distribution, if the covariances reduce to zero, independence follows.

In what follows it will be useful to obtain an expression for the characteristic function (the Fourier transform) of a multivariate normal distribution. By this we mean the integral

$$\varphi(\lambda) = \varphi(\lambda_1, \ldots, \lambda_n)$$

$$= C \int \ldots \int e^{i(\lambda, x) - \frac{1}{2}(A(x-m),(x-m))} dx_1 \ldots dx_n$$

$$= e^{i(\lambda, m)} . C \int \ldots \int e^{i(\lambda, y) - \frac{1}{2}(Ay, y)} dy_1 \ldots dy_n.$$

Again, setting $y = Sz$, we have

$$\varphi(\lambda) = e^{i(\lambda, m)} . C \int \ldots \int e^{i(\lambda, Sz) - \frac{1}{2}(Dz, z)} dz_1 \ldots dz_n$$

$$= e^{i(\lambda, m)} . C \int \ldots \int e^{i(\mu, z) - \frac{1}{2}(Dz, z)} dz_1 \ldots dz_n.$$

Here $\mu = S^* \lambda = S^{-1} \lambda$. The last integral splits into a product of one-dimensional integrals. We use the fact that for any μ and $d > 0$,

$$\frac{1}{\sqrt{2\pi d^{-1}}} \int e^{i\mu z - \frac{1}{2} dz^2} dz = e^{-\frac{\mu^2}{2d}}.$$

This gives

$$C \int \ldots \int e^{i(\mu, z) - \frac{1}{2}(Dz, z)} dz_1 \ldots dz_n = e^{-\frac{1}{2} \sum_i \frac{\mu_i^2}{d_i}} = e^{-\frac{1}{2}(D^{-1}\mu, \mu)}$$

$$= e^{-\frac{1}{2}(SD^{-1}S^{-1}\lambda, \lambda)} = e^{-\frac{1}{2}(A^{-1}\lambda, \lambda)}.$$

Finally we obtain

$$\varphi(\lambda) = e^{i(\lambda, m) - \frac{1}{2}(A^{-1}\lambda, \lambda)}.$$

By using the last expression we can obtain a definition of a more general multivariate normal distribution which is not necessarily non-degenerate. Specifically let $B = A^{-1}$. In the non-degenerate case $A > 0$ and therefore B exists and $B > 0$. A general multivariate normal distribution is a distribution whose characteristic function has the form

$$\varphi(\lambda) = e^{i(\lambda, m) - \frac{1}{2}(B\lambda, \lambda)},$$

where $B \geq 0$ and m is an arbitrary n−dimensional vector. One can show that in the degenerate case the probability distribution is concentrated on a plane of smaller dimension; if we introduce variables on this plane by means of a linear transformation and a translation, then the density expressed in these variables has the same form as before.

Now let $n = n_1 + n_2$. It will be more convenient to use the notation $\xi_1, \ldots, \xi_{n_1}$ and $x_1, \ldots, x_{n_1}$ for the first n_1 random variables and their values, and $\eta_1, \ldots, \eta_{n_2}$ and $y_1, \ldots, y_{n_2}$ for the remaining n_2 random variables and their values. Correspondingly $m = (m^{(1)}, m^{(2)})$, where $m^{(1)}$ $(m^{(2)})$ is the vector of expectations of the random variables ξ_i (η_j), and the matrix A can be written in the form

$$A = \begin{pmatrix} A^{(1)} & B \\ B & A^{(2)} \end{pmatrix}.$$

We now write the density of a non-degenerate normal distribution of the combined random variables ξ_i, η_j in the form

$$p(x_1, \ldots, x_{n_1}; y_1, \ldots, y_{n_2})$$
$$= (2\pi)^{-n/2} \sqrt{\det A} \exp\{-\frac{1}{2}[(A^{(1)}(x - m^{(1)}), (x - m^{(1)}))$$
$$+ 2(B(x - m^{(1)}), (y - m^{(2)})) + (A^{(2)}(y - m^{(2)}), (y - m^{(2)}))]\}.$$

We now find the conditional distribution of the random variables $\xi_1, \ldots \xi_{n_1}$, conditional on fixed values $y_1, \ldots, y_{n_2}$, for the random variables $\eta_1, \ldots, \eta_{n_2}$. We first assume that $m^{(1)} = 0$ and $m^{(2)} = 0$. We have

$$p(x_1, \ldots, x_{n_1}; y_1, \ldots, y_{n_2})$$
$$= (2\pi)^{-n/2} \sqrt{\det A}. \exp\{-\frac{1}{2}[(A^{(1)}x, x) + 2(Bx, y) + (A^{(2)}y, y)]\}$$
$$= (2\pi)^{-n/2} \sqrt{\det A}. \exp\{-\frac{1}{2}(A^{(2)}y, y)\}$$
$$\times \exp\{-\frac{1}{2}[(A^{(1)}(x - Ly), (x - Ly))] + (Bx, y)$$
$$+ \frac{1}{2}(A^{(1)}(x), Ly) + \frac{1}{2}(A^{(1)}Ly, x) - \frac{1}{2}(A^{(1)}Ly, Ly)\}.$$

Here L is an arbitrary $n_2 \times n_1$ matrix, which transforms R^{n_2} into R^{n_1}. Furthermore, by symmetry of $A^{(1)}$, the scalar product $(A^{(1)}Ly, x) = (Ly, A^{(1)}x) = (A^{(1)}x, Ly)$. Therefore

$$(Bx, y) + \frac{1}{2}(A^{(1)}x, Ly) + \frac{1}{2}(A^{(1)}Ly, x)$$
$$= (Bx, y) + (A^{(1)}x, Ly) = (Bx, y) + (L^*A^{(1)}x, y).$$

We now choose L such that $B + L^*A^{(1)} = 0$, i.e.

$$L^* = -B(A^{(1)})^{-1}, \quad L = (-(A^{(1)}))^{-1}B^*.$$

Then for $A^{(3)} = A^{(2)} + L^*A^{(1)}L = A^{(2)} + B(A^*)^{-1}B^*$ we have

$$p(x_1, \ldots, x_{n_1}; y_1, \ldots, y_{n_2}) = (2\pi)^{-n/2}\sqrt{\det A}$$
$$\times \exp\{-\frac{1}{2}(A^{(2)}y, y) - \frac{1}{2}(A^{(1)}Ly, Ly) - \frac{1}{2}[(A^{(1)}(x - Ly), (x - Ly))]\}$$
$$= (2\pi)^{-n_2/2}\sqrt{\det A^{(3)}} \exp\{-\frac{1}{2}(A^{(3)}y, y)\}.(2\pi)^{-n_1/2}$$
$$\times \sqrt{\det A^{(1)}} \exp\{-\frac{1}{2}[(A^{(1)}(x - Ly), (x - Ly))]\}.$$

$$(9.2)$$

The latter expression is the density of a multivariate normal distribution for the random variables $\xi_1, \ldots, \xi_{n_1}$, with vector of expectation Ly and matrix $A^{(1)}$, and

$$p(y_1, \ldots, y_{n_2}) = (2\pi)^{-n_2/2}\sqrt{\det A^{(3)}} e^{-\frac{1}{2}(A^{(3)}y, y)}$$

is the density of a multivariate normal distribution for the random variables $\eta_1, \ldots, \eta_{n_2}$ with zero expectation and matrix $A^{(3)}$. So we have obtained an expression of the type (8.4). In passing we have proved a formula which is well-known in linear algebra,

$$\det A = \det A^{(1)} \cdot \det\left(A^{(2)} + B(A^{(1)})^{-1}B^*\right),$$

and also the inequality $A^{(3)} = A^{(2)} + B(A^{(1)})^{-1}B^* > 0$. The main conclusion from equation (9.2) for probability theory is that the conditional distribution of the random variables $\xi_1, \ldots, \xi_{n_1}$ for fixed values $y_1, \ldots, y_{n_2}$ of the random variables $\eta_1, \ldots, \eta_{n_2}$ is a normal distribution with vector of expectations $Ly = -(A^{(1)})^{-1}B^*y$ and matrix $A^{(1)}$. In other words, the dependence on the conditions manifests itself only in the vector of conditional expectations.

The previous calculations hold for the case where $m^{(1)} = 0$, $m^{(2)} = 0$. In order to obtain the result for arbitrary $m^{(1)}$, $m^{(2)}$ it suffices to replace x, y by $x - m^{(1)}$, $y - m^{(2)}$ in the final answer. We will not go into further details here.

We now return to the original distribution (9.1). Let us assume that $m = 0$. A system of polynomials, which we will call the Hermite polynomials, although as a rule this terminology is in use only for $n = 1$, is closely connected with the density

$$p(x) = (2\pi)^{-n/2}\sqrt{\det A}\, e^{-\frac{1}{2}(Ax,x)}.$$

Let us consider an arbitrary monomial $x_1^{k_1} x_2^{k_2} \ldots x_n^{k_n}$ of degree $k = k_1 + k_2 + \cdots + k_n$. A Hermite polynomial for this monomial is a polynomial $x_1^{k_1} x_2^{k_2} \ldots x_n^{k_n} + Q(x)$ where $Q(x)$ is a polynomial of degree less than k, such that

$$\int \left(x_1^{k_1} x_2^{k_2} \ldots x_n^{k_n} + Q(x)\right)Q_1(x)e^{-\frac{1}{2}(Ax,x)}\,dx = 0$$

for any polynomial $Q_1(x)$ of degree less than k. The following notation if often used for the Hermite polynomial:

$$x_1^{k_1} x_2^{k_2} \ldots x_n^{k_n} + Q(x) =: x_1^{k_1} \ldots x_n^{k_n} :$$

If $P(x)$ is a homogeneous polynomial of degree k, its Hermite polynomial $: P(x) :$ is defined by linearity. The number k is the degree of the Hermite polynomial.

Lemma 9.1.

$$e^{\frac{1}{2}(Ax,x)}\frac{\partial^k}{\partial x_1^{k_1} \ldots \partial x_n^{k_n}}e^{-\frac{1}{2}(Ax,x)} = P_1(x),$$

is a Hermite polynomial of degree k.

Proof. For an arbitrary polynomial Q_1 of degree less than k we have

$$\int P_1(x)Q_1(x)e^{-\frac{1}{2}(Ax,x)}\,dx_1 \ldots dx_n$$

$$= \int \frac{\partial^k}{\partial x_1^{k_1} \ldots \partial x_n^{k_n}}e^{-\frac{1}{2}(Ax,x)}Q_1(x)\,dx_1 \ldots dx_n.$$

We apply k integrations by parts to the last integral; since the degree of Q_1 is less than k we obtain 0.   □

Lecture 10. The Problem of Percolation

The material of this lecture is not on the program of the examination. It is written in a style close to that of a scientific article, and is concerned with a very current and difficult problem, which at the same time can be formulated in a sufficiently elementary way. The hope is that the readers of this lecture can relatively quickly enter this interesting area and pursue their own research.

We begin with the problem of site percolation through the vertices of a two-dimensional lattice. Let V be the set of points (sites) of the plane $x = (x_1, x_2)$ where x_1, x_2 are integers, and $|x_1| \leq L$, $|x_2| \leq L$. We assume that each site of V can be in two states, vacant and occupied, which we denote by 1 and 0. The state of the whole system V is therefore a function $\omega = \{\omega_x\}$ on V, with values in $X = \{1, 0\}$, where ω_x describes the state of the point $x = (x_1, x_2)$.

A path in V is a collection of points $x^{(1)}, \ldots, x^{(r)}$ in V such that $\|x^{(i)} - x^{(i-1)}\| = 1$, $1 < i \leq r$. We say that a percolation with respect to 1 from top to bottom takes place through ω if there exists a path $x^{(1)}, \ldots, x^{(r)}$ such that $x^{(1)} = (x_1^{(1)}, L)$, $x^{(r)} = (x_1^{(r)}, -L)$ and $\omega_{x^{(i)}} = 1$, $1 \leq i \leq r$.

We can define in analogous ways percolation with respect to 0 from top to bottom, and percolation from left to right. Another modification of the percolation problem can be set up in the following way. Let us consider the set V' of edges of V, i.e. the set of pairs of points $b = (x, y)$, $\|x - y\| = 1$, for $x, y \in V$. As before, we assume that each edge, or bond, can be in two states, 1 (passable bond) or 0 (blocked bond). Instead of $\omega = \{\omega_x\}$ we are given $\omega = \{\omega_b\}$, the collection of states for all the bonds. A path is a sequence of bonds $b^{(1)}, \ldots, b^{(r)}$ such that the origin of $b^{(i)}$ is the same as the end of $b^{(i-1)}$, $1 < i \leq r$. We say that percolation (or conductivity) with respect to 1 from top to bottom takes place for ω if there exists a continuous path $b^{(1)}, \ldots, b^{(r)}$ such that $b^{(1)}$ begins on the line $x_2 = L$, $b^{(r)}$ ends on the line $x_2 = -L$, and $\omega_{b^{(i)}} = 1$, $1 \leq i \leq r$.

The bond percolation problem has an immediate physical interpretation. Let us assume that the bonds of the lattice are make of two materials, one conductive and the other non-conductive. Conductivity means that if we put different potentials on the top and bottom sides of V then the current will flow along a conductive path. There is an interesting book by A. L. Efros entitled "Physics and Geometry of Chaos", published in the series "Kvant Library" in

1982 in which a lot of important problems in physics are described which are more or less connected with the property of conductivity.

We denote by Ω the space of all possible ω, and by $\Omega^{(\text{perc})}$ the subset of those $\omega \in \Omega$ for which percolation from top to bottom occurs (in the site or bond percolation problem). We assume, on a given Ω, a probability distribution which corresponds to a sequence of independent identical trials. This means that we are given two numbers, $p = p(1) \geq 0$, $q = 1 - p = p(0) \geq 0$, and that we set $p(\omega) = \prod_x p(\omega_x)$ (in the site percolation case) and $p(\omega) = \prod_b p(\omega_b)$ (in the bond percolation case). Then $P^{(\text{perc})} = P(\Omega^{(\text{perc})}) = \sum_{\omega \in \Omega^{(\text{perc})}} p(\omega)$ is the probability of percolation from top to bottom and it is a function of L and p.

We now say that percolation from top to bottom occurs for a given p if $\lim_{L \to \infty} P^{(\text{perc})} = 1$. The main problem is then to find the structure of the set of p's for which percolation occurs. The conclusion, which is difficult to prove, is that this set is an interval of the form $(p_c, 1]$ and it is not a simple task to find the critical probability p_c beyond which percolation takes place.

From now on we deal with site percolation. Our goal is to prove the following relatively simple statement.

Theorem 10.1. *There exists a $p_c^{(1)} < 1$ such that for all $p \geq p_c^{(1)}$ percolation with respect to 1 from top to bottom and from left to right takes place.*

Proof. Let $\omega = \{\omega_x\} \in \Omega$ and let us make first the following geometric construction. For each $x \in V$, where $\omega_x = 0$, let us construct a closed square $Q(x)$ with smooth corners (see Fig. 10.1), centered at this point, and with sides of length 3.

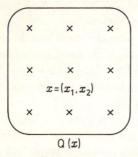

$x = (x_1, x_2)$

$Q(x)$

Fig. 10.1.

Fig. 10.2.

We set $Q(\omega) = \bigcup_{x | \omega_x = 0} Q(x)$ and by convention we call $Q(\omega)$ the domain occupied by 0. Such a domain $Q(\omega)$ can be split into its maximum connected components $Q_1, Q_2, \ldots, Q_r$. We recall that this means that each component Q_i is a connected set, i.e. any two points can be joined by a path which is contained in that set, and that Q_i cannot be enlarged without losing this property. Therefore the Q_i do not intersect. Any two components either lie outside one another, or one component lies in the interior part (the interior) of another component (see Fig. 10.2 where Q is the shaded domain). A component Q_i is said to be exterior if it is not contained in the interior region of another component. Let us denote by $V^{(0)}(\omega)$ the set of points $x = (x_1, x_2)$ which lie outside all exterior components, and therefore outside of all components Q_i, $1 \leq i \leq r$. The following geometric statement is quite clear, and the proof of it will be left to the reader: the set $V^{(0)}(\omega)$ is connected, i.e. any two of its points can be connected by a path. This is due to the fact that $V^{(0}(\omega)$ is obtained by removing the exterior components one after the other, as well as their interior regions, and each such operation preserves connectedness.

It follows from connectedness that percolation from top to bottom with respect to 1 will occur for ω if $V^{(0)}(\omega)$ intersects the lines $x_2 = L$ and $x_2 = -L$. We now show that the probability of each of these events converges to 1 as $L \to \infty$ if p is sufficiently large. It is sufficient to limit ourselves to the case of the line $x_2 = L$. The probability of percolation from left to right is found in the same way.

Let Q be a given connected set. We introduce the indicator function $\chi_Q(\omega)$, which takes the value 1 if Q is one of the components Q_i and 0 otherwise. Consider a point of the form $(x_1, L) \in V$. We will call this point interior (for the configuration ω) if there exists a component Q_i such that (x_1, L) is an element of Q_i or its interior. It is clear that the points which are not interior belong to $V^{(0)}(\omega)$. Furthermore

$$p^{(\text{int})}(x_1) = P\big((x_1, L) \text{ interior }\big) = \sum_{\omega \,|\, (x_1, L) \text{ interior}} p(\omega)$$

$$= \sum_Q \sum_{\omega \in \Omega} p(\omega) \cdot \chi_Q(\omega) = \sum_Q \pi_Q, \text{ with } \pi_Q = \sum_{\omega \in \Omega} p(\omega) \chi_Q(\omega),$$

$$(10.1)$$

where the last sum is carried out over all components Q such that (x_1, L) belongs to Q or to the interior of Q.

Lemma 10.1. *There exist constants $c_1, c_2 < \infty$ such that*

$$\pi_Q \leq (c_1 q)^{c_2 |Q|}.$$

The proof of these lemmas will be carried out a little later. We introduce the lexicographic ordering on the points of the plane. This means that $(x_1, x_2) < (x_1', x_2')$ if either $x_1 < x_1'$ or $x_1 = x_1'$ and $x_2 < x_2'$. Given a connected set Q we define its origin to be the smallest point in the sense of the lexicographic ordering.

Lemma 10.2. *The number of connected sets Q, with $|Q| = n$, which have their origin at the point (x_1, x_2), is not greater than 16^n.*

The proof of this lemma will be carried out a little later. We now find an estimate of $p^{(\text{int})}$. By Lemma 10.1 and (10.1) we have

$$p^{(\text{int})} \leq \sum_{Q} (c_1 q)^{c_2 |Q|},$$

where the sum is carried out over all components Q which contain the point (x_1, L). Furthermore, from Lemma 10.2 we have

$$\sum_{Q} (c_1 q)^{c_2 |Q|} = \sum_{n} \sum_{(x_1, x_2)} \sum_{\substack{Q \mid \text{the origin of } Q \text{ is} \\ (x_1, x_2), |Q| = n}} (c_1 q)^{c_2 |Q|}$$

$$\leq \sum_{n=1}^{\infty} (c_1 q)^{c_2 n} . n^2 . 16^n .$$

Here we used the fact that for a fixed point (x_1, L) and a fixed value $|Q| = n$ the origin of Q can be one of at most n^2 points.

Let p be large enough and $q = 1 - p$ be small enough so that $(c_1 q)^{c_2} 16 < 1$. Then the latter series converges. Let us denote its sum by $d(p)$. It is clear that $d(p) \to 0$ as $p \to 1$.

We introduce the random variable $\xi(\omega)$ equal to the number of interior points on the line (x_1, L), and the random variables $\eta_{x_1}(\omega)$, where

$$\eta_{x_1}(\omega) = \begin{cases} 1, & \text{if } (x_1, L) \text{ is an interior point;} \\ 0 & \text{otherwise.} \end{cases}$$

We clearly have $\xi(\omega) = \sum_{x_1 = -L}^{L} \eta_{x_1}(\omega)$, $E\eta_{x_1}(\omega) = p^{(\text{int})}(x_1)$, and

$$E\xi(\omega) = \sum_{x_1 = -L}^{L} E\eta_{x_1}(\omega) \leq (2L + 1).d(p).$$

This inequality shows that the average number of interior points on the line $x_2 = L$ is relatively small. The latter statement is not sufficient yet to prove the theorem. We also need to estimate the variance $\text{Var}\,\xi$.

We have

$$\text{Var}\,\xi(\omega) = E\left(\sum_{x_1 = -L}^{L} \left(\eta_{x_1}(\omega) - p^{(\text{int})}(x_1) \right)^2 \right)$$

$$= \sum_{x_1, x_2 = -L}^{L} E\left(\eta_{x_1}(\omega) - p^{(\text{int})}(x_1) \right)\left(\eta_{x_2}(\omega) - p^{(\text{int})}(x_2) \right).$$

Furthermore

$$\eta_{x_1}(\omega) = \sum_{Q} \chi_Q(\omega), \quad p^{(\text{int})}(x_1) = \sum_{Q} \pi_Q,$$

where the summation in both cases is carried out over the components Q for which (x_1, L) belongs to Q or to the interior of Q; see 10.1 for the definition of π_Q. Therefore

$$m_{x_1 x_2} = E\big(\eta_{x_1}(\omega) - p^{(\text{int})}(x_1)\big)\big(\eta_{x_2}(\omega) - p^{(\text{int})}(x_2)\big)$$

$$= E\Big(\sum_{Q_{x_1}} (\chi_{Q_{x_1}}(\omega) - \pi_{Q_{x_1}}) \cdot \sum_{Q_{x_2}} (\chi_{Q_{x_2}}(\omega) - \pi_{Q_{x_2}})\Big).$$

Here Q_{x_1} (or Q_{x_2}) denote the components which correspond to x_1 (or x_2). We interchange the operations of summation and expectation and obtain

$$m_{x_1 x_2} = \sum_{Q_{x_1}, Q_{x_2}} E\big(\chi_{Q_{x_1}}(\omega) - \pi_{Q_{x_1}}\big)\big(\chi_{Q_{x_2}}(\omega) - \pi_{Q_{x_2}}\big). \qquad (10.2)$$

We now note, and this is the main probabilistic part of the proof, that the random variables $\chi_{Q_{x_1}} - \pi_{Q_{x_1}}$ and $\chi_{Q_{x_2}} - \pi_{Q_{x_2}}$ are independent if $Q_{x_1} \cap Q_{x_2} = \emptyset$. Therefore

$$E\big(\chi_{Q_{x_1}}(\omega) - \pi_{Q_{x_1}}\big)\big(\chi_{Q_{x_2}}(\omega) - \pi_{Q_{x_2}}\big)$$
$$= E\big(\chi_{Q_{x_1}}(\omega) - \pi_{Q_{x_1}}\big) . E\big(\chi_{Q_{x_2}}(\omega) - \pi_{Q_{x_2}}\big) = 0$$

and therefore in 10.2 we can perform the summation only with respect to those components Q_{x_1} and Q_{x_2} to which both x_1 and x_2 are interior. We denote such a summation by $\sum'$. Then

$$\operatorname{Var}\xi(\omega) = \sum_{x_1, x_2} m_{x_1 x_2} = \sum_{x_1} \sum_{x_2} m_{x_1 x_2}$$

$$= \sum_{x_1} \sum_{Q_{x_1}} \sum_{Q_{x_2}} {}' E\big(\chi_{Q_{x_1}}(\omega) - \pi_{Q_{x_1}}\big)\big(\chi_{Q_{x_2}}(\omega) - \pi_{Q_{x_2}}\big)$$

$$= \sum_{x_1} \sum_{Q_{x_1}} \sum_{Q_{x_2}} {}' \big(E\chi_{Q_{x_1}}(\omega) . \chi_{Q_{x_2}}(\omega) - \pi_{Q_{x_1}} . \pi_{Q_{x_2}}\big).$$

Since two components either agree or do not intersect, $E\chi_{Q_{x_1}}(\omega) . \chi_{Q_{x_2}}(\omega) = 1$ for $Q_{x_1} = Q_{x_2}$ and 0 for $Q_{x_1} \neq Q_{x_2}$. The latter expression therefore takes the form

$$\operatorname{Var}\xi = \sum_{x_1} \sum_{Q_{x_1}} \pi_{Q_{x_1}} (1 - \pi_{Q_{x_1}}) - \sum_{x_1} \sum_{Q_{x_1}} \sum_{Q_{x_2}} {}' \pi_{Q_{x_1}} . \pi_{Q_{x_2}}$$

The first sum is less than or equal to $\sum_{x_1} \sum_{Q_{x_1}} \pi_{Q_{x_1}} = E\xi \leq (2L+1).d(p)$. The second sum is negative. Therefore

$$\operatorname{Var}\xi \leq (2L+1).d(p).$$

We can now use Chebyshev's inequality in its simple form:

$$P\{\xi(x) \leq 2(2L+1).d(p)\}$$
$$= P\{\xi(x) - E\xi(x) \leq 2(2L+1)d(p) - E\xi\}$$
$$\geq P\{\xi(x) - E\xi(x) \leq (2L+1).d(p)\}$$
$$\geq P\Big\{\frac{|\xi(x) - E\xi(x)|}{\sqrt{\operatorname{Var}\xi}} \leq \frac{(2L+1).d(p)}{\sqrt{\operatorname{Var}\xi}}\Big\}$$
$$\geq 1 - \frac{\operatorname{Var}\xi}{(2L+1)^2 . d^2(p)} \geq 1 - \frac{1}{(2L+1)d(p)} \to 1$$

as $L \to \infty$. Let p be large enough so that $2d(p) < 1/2$. Then with probabilities converging to 1 as $L \to \infty$, more than one half of the points on the line $x_2 = L$ are non-interior, i.e. belong to $V^{(0)}(\omega)$. $\qquad\square$

Lemmas 10.1 and 10.2 must still be proved.

Proof of Lemma 10.1. Assume that a set Q is given and that values ω_x are fixed for $x \in Q$. The probability of such an event is equal to $p^k q^{|Q|-k}$, where k is the number of ones, and $|Q| - k$ is the number of zeroes. We now note that for a given set Q the number of subsets of Q where a zero appears is less than or equal to the total number of all subsets of Q, i.e. $2^{|Q|}$. Furthermore, near each one, at a distance less than or equal to $\sqrt{2}$, lies a 0, since Q is the union of squares (such as in Fig. 10.1) at the center of each of which lies a 0. Therefore $|Q| - k \geq (1/9)|Q|$. Finally we obtain

$$\pi(Q) \leq 2^{|Q|} \cdot q^{(1/9)|Q|} = (c_1 q)^{(1/9)|Q|}.$$

But this is the statement of the lemma for $c_1 = 2^9$, $c_2 = 1/9$. $\qquad\square$

Proof of Lemma 10.2. We show that to each connected set Q we can associate a word $(\epsilon_1, \epsilon_2, \ldots, \epsilon_n)$ of length n, where ϵ_i takes at most 16 values, in such a way that Q is uniquely determined by this word.

Let x be a point in Q. We consider the set R of its neighbors, i.e. those points y such that $d(x, y) \leq 1$ and $y \in Q$. The number of such subsets is no greater than the number of all subsets of a set with four elements (the set of all neighbors), which equals $2^4 = 16$, so R can be indexed by numbers from 1 to 16.

We now consider the point $x = (x_1, x_2)$ which is the origin of Q, and the set R_1 of its neighbors which belong to Q. We then consider the smallest point in R_1 which is distinct from x, and we denote by R_2 the set of its neighbors. Then we consider the smallest of the remaining points, and denote by R_3 the set of its neighbors, and so forth. In this fashion we obtain a word $(R_1, R_2, \ldots, R_n)$ of length n in which each R_i takes at most 16 values. It is clear that Q is uniquely determined by this word. $\qquad\square$

For the bond percolation problem the exact value of the critical probability is known to be equal to $1/2$. This was proved rigorously relatively recently by the American mathematician H. Kesten (see H. Kesten, "Percolation for Mathematicians", Birkhäuser, 1982). Many numerical experiments show that for the site percolation problem for a two-dimensional square lattice, $p_c \simeq 0.58$. The Japanese mathematician I. Higuchi showed that $p_c > 1/2$. Recently a Hungarian mathematician, B. Tot, proved by elementary methods that $p_c > 0.503$. We also quote the result of the Soviet mathematician L. Mitiushin, which shows that in the case of the lattice $\mathbb{Z}^d$ in a space of sufficiently high dimension d, for the site percolation problem, there exists an interval of values p which contains the value $p = 1/2$ in its interior, for which percolation is possible both with respect to 1 and with respect to 0.

Lecture 11. Distribution Functions, Lebesgue Integrals and Mathematical Expectation

11.1 Introduction

In this lecture we introduce once more the fundamental concepts of probability theory, which we have encountered already, but in a much more general form than previously.

We define a probability space as a triplet $(\Omega, \mathcal{F}, P)$ where Ω is the space of elementary outcomes, $\mathcal{F}$ is a σ-algebra of subsets of Ω and P is a probability measure defined on the events $C \in \mathcal{F}$.

The choice of the σ-algebra $\mathcal{F}$ plays an important role in many problems, in particular in those problems which are connected to the theory of stochastic processes. In the cases where Ω is a complete separable metric space it is usual to take for $\mathcal{F}$ the Borel σ-algebra, that is, the smallest σ-algebra which contains all the open balls, i.e. the sets of the form $\{\omega | \rho(\omega, \omega_0) < r\} = B_r(\omega)$. Such a σ-algebra is denoted by $\mathcal{B}(\Omega)$.

Let $\xi = f(\omega)$ be a real-valued function on Ω.

Definition 11.1. ξ is said to be a random variable if for any Borel set $A \subset \mathbb{R}^1$

$$\{\omega | f(\omega) \in A\} = f^{-1}(A) \in \mathcal{F}.$$

The property given in Definition 11.1 is sometimes called the measurability of f with respect to the σ-algebra $\mathcal{F}$.

It is easy to check that

1) $f^{-1}(\mathbb{R}^1) = \Omega$;
2) $f^{-1}(\mathbb{R}^1 \setminus A) = \Omega \setminus f^{-1}(A)$;
3) $f^{-1}(\bigcup_{i=1}^{\infty} A_i) = \bigcup_{i=1}^{\infty} f^{-1}(A_i)$.

It follows that the class of subsets of the form $f^{-1}(A)$, where $A \in \mathcal{B}(\mathbb{R}^1)$, is a σ-subalgebra of the σ-algebra $\mathcal{F}$. This σ-algebra is said to be the σ-algebra generated by the random variable ξ. We will denote it by $\mathcal{F}_\xi$.

We assume that the random variable ξ takes a finite or countable number of values which we write in a sequence $\{a_1, a_2, \ldots\}$. We set $C_i = \{\omega | \xi = f(\omega) = a_i\}$. Then

1) $C_i \cap C_j = \emptyset$ for $i \neq j$,
2) $\bigcup_i C_i = \Omega$.

In other words, the events C_i define a partition of the space Ω. It is easy to see that each $C \in \mathcal{F}_\xi$ is a union of some C_i. One can formally construct the partition which corresponds to an arbitrary random variable ξ. For this we introduce the sets $C_t = \{\omega | f(\omega) = t\}$, $-\infty < t < \infty$. It is clear that the C_t, for distinct t, do not intersect and that $\bigcup_{-\infty < t < \infty} C_t = \Omega$. One can show, under some additional assumptions on the probability space $(\Omega, \mathcal{F}, P)$, that this partition is measurable (see Lecture 8). The corresponding probability distribution on C_t is called the conditional distribution, conditional on the random variable ξ taking on the value t.

We set $P_\xi(\mathcal{A}) = P(f^{-1}(\mathcal{A}))$ for every $\mathcal{A} \in \mathcal{B}(\mathbb{R}^1)$. It is easy to check that $P_\xi(\mathcal{A})$ is a probability measure defined on the Borel σ-algebra $\mathcal{B}(\mathbb{R}^1)$.

Definition 11.2. P_ξ is called the probability distribution of the random variable ξ.

We define $\mathcal{A}_t = (-\infty, t] = \{x \in \mathbb{R}^1, x \leq t\}$.

Definition 11.3. Let P be a probability measure on the line $\mathbb{R}^1$. The distribution function F of the measure P is a function of the variable t, $-\infty < t < \infty$, given by the relation $F(t) = P(\mathcal{A}_t)$. If $P = P_\xi$ then $F_\xi(t) = P_\xi(\mathcal{A}_t)$ is called the distribution function of the random variable ξ.

11.2 Properties of Distribution Functions

1) $F_\xi(t_1) \leq F_\xi(t_2)$ for $t_1 < t_2$. Indeed $\mathcal{A}_{t_1} \subset \mathcal{A}_{t_2}$ for $t_1 < t_2$ and therefore $f^{-1}(\mathcal{A}_{t_1}) \subseteq f^{-1}(\mathcal{A}_{t_2})$. It then follows from the properties of probabilities that $P(f^{-1}(\mathcal{A}_{t_1})) \leq P(f^{-1}(\mathcal{A}_{t_2}))$.

2) $\lim_{t \to \infty} F_\xi(t) = 1$, $\lim_{t \to -\infty} F_\xi(t) = 0$. We only prove the first statement as the second statement is proved in the same way. Let us take a monotone sequence $\{t_i\}$, with $t_i \to \infty$ as $i \to \infty$. Then the $\{\mathcal{A}_{t_i}\}$ form a monotonically increasing sequence of subsets of the line and $\bigcup_i \mathcal{A}_{t_i} = \mathbb{R}^1$. Therefore the $f^{-1}(\mathcal{A}_{t_i})$ form a monotonically increasing sequence of subsets of Ω with $\bigcup_i f^{-1}(\mathcal{A}_{t_i}) = \Omega$. It follows from the σ-additivity of the probability P that

$$\lim_{i \to \infty} F_\xi(t_i) = \lim_{i \to \infty} P_\xi(\mathcal{A}_{t_i}) = \lim_{i \to \infty} P(f^{-1}(\mathcal{A}_{t_i})) = 1.$$

3) $F_\xi(t)$ is continuous from the right. Indeed let $t_i \downarrow \bar{t}$. Then the $\mathcal{A}_{t_i}$ decrease and $\bigcap_i \mathcal{A}_{t_i} = \mathcal{A}_{\bar{t}}$. Therefore from the σ-additivity of probabilities we have

$$\lim_{i \to \infty} F_\xi(t_i) = \lim_{i \to \infty} P(f^{-1}(\mathcal{A}_{t_i})) = P(f^{-1}(\mathcal{A}_{\bar{t}})) = F_\xi(\bar{t}).$$

4) For any $\bar{t}$ the probability $P(\xi = \bar{t}) = F_\xi(\bar{t}) - \lim_{t_i \to \bar{t} - 0} F_\xi(t_i)$ for any sequence $\{t_i\}$, $t_i \uparrow \bar{t}$. Indeed $\bigcup_i \mathcal{A}_{t_i} = (-\infty, \bar{t})$. Therefore

$$\lim_{i \to \infty} F_\xi(t_i) = P\{\xi < \bar{t}\},$$

$$F_\xi(\bar{t}) - \lim_{t_i \to \bar{t}-0} F_\xi(t_i) = P(\xi \le \bar{t}) - P(\xi < \bar{t}) = P(\xi = \bar{t}).$$

It follows from Property 4) that $P(\xi = \bar{t}) = 0$ if and only if F_ξ is continuous at the point $\bar{t}$.

For any half-open interval $(s_1, s_2]$ with $s_1 < s_2$, we have

$$P_\xi((s_1, s_2]) = P(\{\omega | s_1 < \xi \le s_2\})$$
$$= P(\{\omega | \xi \le s_2\}) - P(\{\omega | \xi \le s_1\}) = F_\xi(s_2) - F_\xi(s_1).$$

Thus the distribution function $F_\xi(t)$ uniquely determines the values of P_ξ on the half-open intervals $(s_1, s_2]$. It follows from the general theory of extensions of measures that any σ-additive probability measure, defined on the Borel σ-algebra on the line, is uniquely determined by its values on such half-open intervals $(s_1, s_2]$ (Caratheodory's Theorem). Consequently the distribution function $F_\xi(t)$ uniquely determines the probability distribution P_ξ.

11.3 Types of Distribution Functions

Discrete Type. Assume that the random variable ξ takes a finite or countable number of values $a_1, a_2, a_3, \ldots$ with probabilities $p_1, p_2, p_3, \ldots$, respectively. By definition $F_\xi(t) = \sum_{i: a_i \le t} p_i$. Such a distribution function is a step function. The jumps occur at the points a_i and the height of the jumps is equal to p_i.

Absolutely Continuous Type (with Respect to the Lebesgue Measure). This type is that of a distribution function $F_\xi(t)$ for which there exists a non-negative function $p_\xi(t)$ such that $F_\xi(t) = \int_{-\infty}^{t} p_\xi(u)\, du$. In such a case $p_\xi(t)$ is called the density of the distribution. We already encountered several examples of densities in earlier lectures. It is clear that $F_\xi(t)$ is continuous at every t.

Singular Type (with Respect to the Lebesgue Measure). There are continuous distribution functions $F_\xi(t)$ which cannot be represented in the form of an integral with respect to a density.

The so-called Cantor Staircase is an example of such a situation. We set $F_\xi(t) = 1/2$ for $1/3 \le t \le 2/3$, $F_\xi(t) = 1/4$ for $1/9 \le t \le 2/9$, and $F_\xi(t) = 3/4$ for $7/9 \le t \le 8/9$. The construction process can be described inductively in the following way. At the nth step we have intervals of length 3^{-n}, where the function $F_\xi(t)$ is not yet defined, although it is defined at the end-points of such intervals. Let us divide any such interval into three parts, and set $F_\xi(t)$ to be constant on the middle interval, and equal to the half-sum of its values at the end-points of the preceding interval. For all remaining t the function $F_\xi(t)$ is extended by continuity. The limit function is called the Cantor Staircase (sometimes the Devil's Staircase). Singular distribution functions arise more and more often in applications.

Any distribution function can be represented in a unique way as a sum of distribution functions of the three types above.

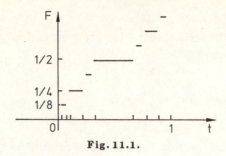

Fig. 11.1.

In what follows, when talking about distribution functions we will always mean functions $F(x)$ which satisfy Properties 1) - 3), without referring to their dependence on ξ.

The random variables which we introduced in Definition 11.1 are sometimes called real valued random variables. It is often useful to consider complex or vector-valued random variables or, more generally, random variables with values in an abstract space. We now introduce a corresponding definition. Let $(X, \mathcal{X})$ be a measurable space.

Definition 11.4. The transformation $\xi = f(\omega) : \Omega \to X$ is said to be measurable if for any $A \in \mathcal{X}$ the inverse image $f^{-1}(A) = \{\omega | f(\omega) \in A\} \in \mathcal{F}$.

The transformation ξ will be called an X-valued random variable. We set $P_\xi(A) = P(\{\omega | f(\omega) \in A\})$ for any $A \in \mathcal{X}$. Then P_ξ is a probability measure defined on the σ-algebra $\mathcal{X}$. It is called the probability distribution corresponding to the transformation ξ.

We now return to real valued random variables. A random variable is said to be simple if it takes a finite or countable number of values. Simple random variables form a vector space over the field of real numbers. The product of a finite number of simple random variables is a simple random variable. The quotient of two simple random variables is a simple random variable, provided that the denominator does not take the value zero.

Theorem 11.1. *Any random variable* $\xi = f(\omega)$ *is a monotone limit from below of simple random variables, i.e.* $\xi = f(\omega) = \lim_{n \to \infty} f_n(\omega)$, *where* $\xi_n = f_n(\omega)$ *are simple random variables and* $f_n(\omega) \uparrow f(\omega)$ *for every* ω. *Conversely, if* $\xi = f(\omega)$ *is a monotone limit from below of simple random variables then* ξ *is also a random variable.*

Proof. Let $\xi_n = f_n(\omega)$ be defined by the relations

$$f_n(\omega) = k2^{-n} \quad \text{if} \quad k2^{-n} < f(\omega) \le (k+1)2^{-n}.$$

The sequence ξ_n satisfies the requirements of the theorem.

We now prove the converse statement. Given a function $\xi = f(\omega)$ we consider the subsets $A \subset \mathbb{R}^1$ for which $f^{-1}(A) \in \mathcal{F}$. It is easy to see that the class of such subsets forms a σ-algebra which we will denote by $\mathcal{R}_\xi$. We prove that the half-open intervals $A_t = (-\infty, t] \in \mathcal{R}_\xi$. Indeed it is easy to check the following relation:

$$f^{-1}(A_t) = \bigcap_k \bigcup_n \bigcap_{m \geq n} \{\omega | f_m(\omega) \leq t + 1/k\}.$$

Since $\{\omega | f_n(\omega) \leq t - 1/k\} \in \mathcal{F}$, $f^{-1}(A_t) \in \mathcal{F}$, i.e. $A_t \in \mathcal{R}_\xi$. In Lecture 2 we proved that the smallest σ-algebra which contains all A_t is $\mathcal{B}(\mathbb{R}^1)$. Therefore $\mathcal{B}(\mathbb{R}^1) \subset \mathcal{R}_\xi$. $\qquad\square$

Let $g(x_1, \ldots, x_p)$ be a function of p real variables which is measurable with respect to the σ-algebra $\mathcal{B}(\mathbb{R}^p)$.

Theorem 11.2. *For any random variables $\xi_1 = f_1(\omega), \ldots, \xi_p = f_p(\omega)$ the function $\eta = g(\xi_1, \ldots, \xi_p)$ is also a random variable.*

Proof. Using Theorem 11.1 we construct a monotone sequence of simple functions $g_n(x_1, \ldots, x_p)$ such that $g_n \uparrow g$ for any choice of $(x_1, \ldots, x_p)$. Then η is a monotone limit of the functions $\eta_n = g_n(\xi_1, \ldots, \xi_p)$, which take at most a countable number of values. It is therefore enough to prove that η_n is a simple random variable.

We denote the values of η_n by $a_1, a_2, \ldots$, and we set $A_i \in \mathbb{R}^p = \{(x_1, \ldots, x_p) \in \mathbb{R}^p | g_n(x_1, \ldots, x_p) = a_i\}$. Then $A_i \in \mathcal{B}(\mathbb{R}^p)$ since the g_n are measurable with respect to the σ-algebra $\mathcal{B}(\mathbb{R}^p)$. To complete the proof of the theorem we must prove that for any $A \in \mathcal{B}(\mathbb{R}^p)$ we have $\{\omega | (f_1(\omega), \ldots, f_p(\omega)) \in A\} \in \mathcal{F}$. We use the following statement: let $A_{t_1, \ldots, t_p} = \{(x_1, \ldots, x_p) : x_i \leq t_i \text{ for } 1 \leq i \leq p\}$; then $\mathcal{B}(\mathbb{R}^p)$ is the smallest σ-algebra generated by all sets of the form $A_{t_1, \ldots, t_p}$. The proof of this statement is carried out in exactly the same way as for $p = 1$.

We now argue as in the proof of Theorem 11.1. Denote by $\mathcal{R}_{\xi_1, \ldots, \xi_p}$ the class of those subsets $A \subset \mathbb{R}^p$ such that $\{\omega | (f_1(\omega), \ldots, f_p(\omega)) \in A\} \in \mathcal{F}$. This class forms a σ-algebra of subsets of $\mathbb{R}^p$. Also $A_{t_1, \ldots, t_p} \in \mathcal{R}_{\xi_1, \ldots, \xi_p}$ since

$$\{\omega | (f_1(\omega), \ldots, f_p(\omega)) \in A_{t_1, \ldots, t_p}\} = \bigcap_{i=1}^p \{\omega | f_i(\omega) \leq t_i\} \in \mathcal{R}_{\xi_1, \ldots, \xi_p}.$$

Consequently $\mathcal{B}(\mathbb{R}^p) \subseteq \mathcal{R}_{\xi_1, \ldots, \xi_p}$ and therefore

$$\{\omega | (f_1(\omega), \ldots, f_p(\omega)) \in A_i\} \in \mathcal{F}, \quad i = 1, 2, \ldots$$

Thus $\eta_n = g_n(\xi_1, \ldots, \xi_p)$ is a random variable. $\qquad\square$

Corollary 11.1. *If $\xi_1, \ldots, \xi_p$ are random variables and $a_1, \ldots, a_p$ are real numbers, then $\eta = a_1 \xi_1 + \cdots + a_p \xi_p$ is a random variable.*

Corollary 11.2. *If $\xi_1, \ldots, \xi_p$ are random variables then $\eta = \xi_1 \times \cdots \times \xi_p$ is also a random variable.*

Corollary 11.3 *If ξ_1 and ξ_2 are random variables, with $\xi_2 \neq 0$ for every ω then $\eta = \xi_1 / \xi_2$ is also a random variable.*

In order to prove Corollary 11.1 it suffices to take as g in Theorem 11.2 $g(x_1, \ldots, x_p) = a_1 x_1 + \cdots a_p x_p$. To prove Corollary 11.2 we just need to take $g(x_1, \ldots, x_p) = x_1 \times \cdots \times x_p$. And to prove Corollary 11.3 we take $g(x_1, x_2) = x_1/x_2$.

The Lebesgue integral plays a central role in probability theory. We briefly recall how it is constructed. Let $\xi = f(\omega)$ be a simple random variable taking non-negative values, which we denote by $a_1, a_2, \ldots$ and let $C_i = \{\omega | f(\omega) = a_i\}$.

Definition 11.5. The series $E\xi = \sum_{i=1}^{\infty} a_i P(C_i)$ is said to be the Lebesgue integral of the random variable ξ.

In probability theory $E\xi$ is called the mathematical expectation of the random variable ξ. If $E\xi < \infty$ we say that the random variable ξ has a finite mathematical expectation. It is clear that

1) $E\xi \geq 0$;
2) $E\mathbb{1} = 1$ where $\mathbb{1}$ is the random variable identically equal to 1 on Ω;
3) $E(a\xi_1 + b\xi_2) = aE\xi_1 + bE\xi_2$ for any $a, b > 0$;
4) $E\xi_1 \geq E\xi_2$ if $\xi_1 \geq \xi_2$.

We give a short proof for Property 4). Let the values of the random variable ξ_i be denoted by $a_1^{(i)}, a_2^{(i)}, \ldots$ for $i = 1, 2$, and let $A_j^{(i)}$ denote the set on which the value $a_j^{(i)}$ is taken. We set $B_{j_1, j_2} = A_{j_1}^{(1)} \cap A_{j_2}^{(2)}$. Then on the set B_{j_1, j_2}, ξ_1 takes the value a_{j_1} and ξ_2 takes the value a_{j_2}. We now have

$$E\xi_1 = \sum_{j_1} a_{j_1}^{(1)} P(A_{j_1}^{(1)}) = \sum_{j_1, j_2} a_{j_1}^{(1)} P(B_{j_1, j_2}),$$

$$E\xi_2 = \sum_{j_2} a_{j_2}^{(2)} P(A_{j_2}^{(2)}) = \sum_{j_1, j_2} a_{j_2}^{(2)} P(B_{j_1, j_2}),$$

and

$$E\xi_1 - E\xi_2 = \sum_{j_1, j_2} (a_{j_1}^{(1)} - a_{j_2}^{(2)}) P(B_{j_1, j_2}) \geq 0,$$

since $a_{j_1}^{(1)} \geq a_{j_2}^{(2)}$ on any B_{j_1, j_2}.

Now let $\xi = f(\omega)$ be an arbitrary random variable taking non-negative values. We consider the sequence $\xi_n = f_n(\omega)$ of non-negative simple random variables which converge monotonically to $\xi = f(\omega)$ from below, i.e. $\xi_{n_1} \geq \xi_{n_2}$ for $n_1 \geq n_2$ and $f(\omega) = \lim_{n \to \infty} f_n(\omega)$ for every ω. It follows from Property

4) of $E\xi$ that the sequence of the $E\xi_n$ is non-decreasing and therefore that there exists a limit $\lim_{n \to \infty} E\xi_n$ which is possibly infinite.

Theorem 11.3. *The value of* $\lim_{n \to \infty} E\xi_n$ *does not depend on the choice of the approximating sequence.*

Proof. We first establish the following lemma. Let $\eta \geq 0$ be a simple random variable, $\eta \leq \xi$, and assume that $\xi = \lim_{n \to \infty} \xi_n$, where the ξ_n are non-negative simple random variables, $\xi_{n+1} \geq \xi_n$.

Lemma 11.1. $E\eta \leq \lim_{n \to \infty} E\xi_n$.

Proof. Let us take an arbitrary $\epsilon > 0$ and set $C_n = \{\omega | \xi_n(\omega) - \eta(\omega) > -\epsilon\}$. It then follows from the monotonicity of ξ_n that $C_n \subseteq C_{n+1}$ and it follows from the fact that $\xi_n \uparrow \xi$, and from the inequality $\xi \geq \eta$, that $\bigcup_n C_n = \Omega$. Therefore $P(C_n) \to 1$ as $n \to \infty$. We introduce the indicator function χ_{C_n} where

$$\chi_{C_n}(\omega) = \begin{cases} 1 & \text{for } \omega \in C_n, \\ 0 & \text{for } \omega \notin C_n. \end{cases}$$

Then $\eta_n = \eta \cdot \chi_{C_n}$ is a simple random variable and $\eta_m \leq \xi_m(\omega) + \epsilon$. Therefore by the monotonicity of $E\xi_m$ we have

$$E\eta_k \leq E\xi_k(\omega) + \epsilon, \quad E\eta_k \leq \lim_{m \to \infty} E\xi_m + \epsilon.$$

Since ϵ was arbitrary we obtain $E\eta_k \leq \lim_{m \to \infty} E\xi_m$. We still must prove that $\lim_{k \to \infty} E\eta_k = E\eta$.

We denote by $b_1, b_2, \ldots$ the values of the random variable η and by B_i the set where the value b_i is taken, $i = 1, 2, \ldots$. Then

$$E\eta = \sum_i b_i P(B_i), \quad E\eta_k = \sum_i b_i P(B_i \cap C_k).$$

It is clear that for all i we have $\lim_{k \to \infty} P(B_i \cap C_k) = P(B_i)$.

Since the series above consist of non-negative terms, and since the convergence is monotonous for each i, we have

$$\lim_{k \to \infty} E\eta_k = \lim_{k \to \infty} \sum_i b_i P(B_i \cap C_k)$$
$$= \sum_i b_i \lim_{k \to \infty} P(B_i \cap C_k) = \sum_i b_i P(B_i) = E\eta.$$

It is now easy to obtain the independence of $\lim_{n \to \infty} E\xi_n$ from the choice of the approximating sequence.

Let there be two sequences $\{\xi_n^{(1)}\}$, $\xi_{n+1}^{(1)} \geq \xi_n^{(1)}$ and $\{\xi_n^{(2)}\}$, $\xi_{n+1}^{(2)} \geq \xi_n^{(2)}$ such that $\lim_{n \to \infty} \xi_n^{(1)} = \lim_{n \to \infty} \xi_n^{(2)} = \xi$ for every ω. It follows from Lemma 11.1 that for any k, $E\xi_k^{(1)} \leq \lim_{n \to \infty} E\xi_n^{(2)}$ and therefore $\lim_{k \to \infty} E\xi_k^{(1)} \leq$

$\lim_{n \to \infty} E\xi_n^{(2)}$. We obtain $\lim_{n \to \infty} E\xi_n^{(2)} \leq \lim_{k \to \infty} E\xi_k^{(1)}$, by exchanging $\xi_k^{(1)}$ and $\xi_k^{(2)}$, i.e. $\lim_{n \to \infty} E\xi_n^{(1)} = \lim_{n \to \infty} E\xi_n^{(2)}$.

The limit $\lim_{n \to \infty} E\xi_n$ is called the mathematical expectation of the random variable ξ. We will denote it by the symbol $E\xi$.

The mathematical expectation thus defined satisfies Properties 1) - 4) given above.

Now let ξ be an arbitrary random variable. We introduce the indicator functions:

$$\chi_+(\omega) = \begin{cases} 1, & \text{if } \xi = f(\omega) \geq 0, \\ 0, & \text{if } \xi = f(\omega) < 0, \end{cases}$$

$$\chi_-(\omega) = \begin{cases} 1, & \text{if } \xi = f(\omega) < 0, \\ 0, & \text{if } \xi = f(\omega) \geq 0. \end{cases}$$

Then $\chi_+(\omega) + \chi_-(\omega) \equiv 1$, $\xi = \xi\chi_+ + \xi\chi_- = \xi_1 - \xi_2$, where $\xi_1 = \xi\chi_+$ and $\xi_2 = -\xi\chi_-$. Moreover $\xi_1 \geq 0$, $\xi_2 \geq 0$ so the mathematical expectations $E\xi_1$ and $E\xi_2$ have already been defined.

Definition 11.6. The random variable ξ has a finite mathematical expectation if $E\xi_1 < \infty$ and $E\xi_2 < \infty$. In this case it is equal to $E\xi = E\xi_1 - E\xi_2$. If $E\xi_1 = \infty$ and $E\xi_2 < \infty$ ($E\xi_1 < \infty$, $E\xi_2 = \infty$) then $E\xi = \infty$ ($E\xi = -\infty$). If $E\xi_1 = \infty$ and $E\xi_2 = \infty$ then $E\xi$ is not defined.

Since $|\xi| = \xi_1 + \xi_2$, $E|\xi| = E\xi_1 + E\xi_2$ and so $E\xi$ is finite if and only if $E|\xi|$ is finite.

The mathematical expectation $E\xi$ that we have introduced satisfies all properties described in Lecture 1. In particular

1) $E(a_1\xi_1 + a_2\xi_2) = a_1 E\xi_1 + a_2 E\xi_2$ if $E\xi_1$ and $E\xi_2$ are finite;
2) $E\xi \geq 0$ if $\xi \geq 0$;
3) $E\mathbb{1} = 1$.

One sometimes says that $E\xi$ is a non-negative normed linear functional on the linear space of random variables. The space of random variables ξ for which $E|\xi| < \infty$ is denoted by $\mathcal{L}^1(\Omega, \mathcal{F}, P)$.

The variance of the random variable ξ is defined to be $E(\xi - E\xi)^2$, the n-th order moment is defined to be $E\xi^n$, and the n-th order centralized moment is defined to be $E(\xi - E\xi)^n$. Given two random variables ξ_1 and ξ_2 their covariance is defined by $\text{Cov}(\xi_1, \xi_2) = E(\xi_1 - E\xi_1)(\xi_2 - E\xi_2)$. The correlation coefficient of two random variables ξ_1, ξ_2 is defined to be $\rho(\xi_1, \xi_2) = E(\xi_1 - E\xi_1)(\xi_2 - E\xi_2)/\sqrt{\text{Var}\xi_1 \times \text{Var}\xi_2}$.

The space of random variables ξ for which $E|\xi|^p < \infty$ is denoted by $\mathcal{L}^p(\Omega, \mathcal{F}, P)$.

It is an essential fact in probability theory that the mathematical expectation, the moments, the variance and so forth of a random variable ξ can be found if the measure P_ξ is known or if the distribution function F_ξ is known.

Theorem 11.4. *Let ξ be a random variable and $g(x)$ be a Borel function on $\mathbb{R}^1$. Then for the random variable $\eta = g(\xi)$*

$$En = \int_{-\infty}^{\infty} g(x)\, dP_\xi(x).$$

If P_ξ has a density $p_\xi(x)$ then the latter integral can be written as a Lebesgue integral $En = \int_{-\infty}^{\infty} g(x)p_\xi(x)\, dx$.

Corollary 11.4 $E\xi = \int_{-\infty}^{\infty} x\, dP_\xi(x)$.

Corollary 11.5 $E\xi^p = \int_{-\infty}^{\infty} x^p\, dP_\xi(x)$.

Corollary 11.6 $\mathrm{Var}\xi = \int_{-\infty}^{\infty} (x - E\xi)^2\, dP_\xi(x)$.

Proof of theorem. As usual we first consider the case when $g \geq 0$ and takes a finite or countable number of values $g_1, g_2, \ldots$. Let $C_i = \{x | g(x) = g_i\} \in B(\mathbb{R}^1)$, $A_i = g^{-1}(C_i) \in \mathcal{F}$. Then the random variable η takes the values $g_1, g_2, \ldots$ and by definition $E\eta = \sum g_i P(C_i)$. On the other hand, also by definition, $P(C_i) = P_\xi(A_i)$ and $\sum g_i P_\xi(A_i) = \int g(x)\, dP_\xi(x)$. Therefore the theorem is proved for simple non-negative g. For an arbitrary non-negative g the result follows by passage to the limit in a sequence of simple non-negative functions. In the general case we write $g = g_1 - g_2$ where $g_1, g_2 \geq 0$ and we use the statement of the theorem separately for g_1 and g_2. $\qquad\square$

Given several random variables $\xi_1, \ldots \xi_p$ the concept of joint distribution function is often used: $F_{\xi_1, \ldots, \xi_p}(x_1, \ldots, x_p) = P\{\xi_1 \leq x_1, \ldots, \xi_p \leq x_p\}$. As in the case where $p = 1$, the function $F_{\xi_1, \ldots, \xi_p}$ uniquely defines the measure $P_{\xi_1, \ldots, \xi_p}$ on $B(\mathbb{R}^p)$, where $P_{\xi_1, \ldots, \xi_p}(A) = P(\{\omega | (\xi_1(\omega), \ldots, \xi_p(\omega)) \in A\})$, $A \in B(\mathbb{R}^p)$. The joint distribution function uniquely determines the joint distribution function of any subset of the set of random variables $(\xi_1, \ldots, \xi_p)$ and for the random variable $\eta = g(\xi_1, \ldots, \xi_p)$

$$En = \int \cdots \int_{\mathbb{R}^p} g(x_1, \ldots, x_p)\, dP_{\xi_1, \ldots, \xi_p}(x_1, \ldots, x_p).$$

Lecture 12. General Definition of Independent Random Variables and Laws of Large Numbers

The definition of independent real-valued random variables was given in Lecture 4. We now extend this definition to the case of more general random variables.

Let $(X, \mathcal{X})$ be a measurable space, and ξ_1, ξ_2 be two X-valued random variables, i.e. $\xi_1 = f_1(\omega)$, $\xi_2 = f_2(\omega)$ are given as measurable transformations from $\Omega \to X$. Then by analogy with Lecture 4 the random variables ξ_1, ξ_2 are said to be independent if for any $\mathcal{A}_1, \mathcal{A}_2 \in \mathcal{X}$ we have the equality: $P\{\xi_1 \in \mathcal{A}_1, \xi_2 \in \mathcal{A}_2\} = P(\{\omega | f_1(\omega) \in \mathcal{A}_1, f_2(\omega) \in \mathcal{A}_2\}) = P\{\xi_1 \in \mathcal{A}_1\}P\{\xi_2 \in \mathcal{A}_2\}$. In the general case let $\xi_1 = f_1(\omega)$, $\xi_2 = f_2(\omega), \ldots, \xi_n = f_n(\omega)$ be a collection of n X-valued random variables.

Definition 12.1. The random variables $\xi_1, \xi_2, \ldots, \xi_n$ are said to be jointly independent if for any $\mathcal{A}_1, \mathcal{A}_2, \ldots, \mathcal{A}_n \in \mathcal{X}$

$$P\{\xi_i \in \mathcal{A}_i, 1 \leq i \leq n\} = P(\{\omega | f_i(\omega) \in \mathcal{A}_i, 1 \leq i \leq n\}) = \prod_{i=1}^{n} P\{\xi_i \in \mathcal{A}_i\}.$$

We now prove three simple statements about the joint independence of random variables.

Lemma 12.1. *Let $g_1(x), \ldots, g_n(x)$ be real-valued measurable functions on X. Then the random variables $\eta_1 = g_1(\xi_1), \eta_2 = g_2(\xi_2), \ldots, \eta_n = g_n(\xi_n)$ are jointly independent.*

Proof. Let $\mathcal{A}_i$, $1 \leq i \leq n$, be Borel subsets of the line and let $\mathcal{A}'_i \in \mathcal{X}$ be defined as $g_i^{-1}(\mathcal{A}_i)$, i.e. $\mathcal{A}'_i = \{x \in X | g_i(x) \in \mathcal{A}_i\}$. Then

$$P\{\eta_i \in \mathcal{A}_i, 1 \leq i \leq n\} = P\{\xi_i \in \mathcal{A}'_i, 1 \leq i \leq n\}$$

$$= \prod_{i=1}^{n} P\{\xi_i \in \mathcal{A}'_i\} = \prod_{i=1}^{n} P\{\eta_i \in \mathcal{A}_i\}.$$

In these equalities we used the joint independence of the ξ_i, $1 \leq i \leq n$. □

Lemma 12.2. *Under the conditions of Lemma 12.1, assume that the random variables $\eta_1, \ldots, \eta_m$ are such that $|E\eta_i| < \infty$, $1 \leq i \leq m$. Then the random* *variable* $\eta = \eta_1 \times \cdots \times \eta_m$ *has a finite mathematical expectation and* $E\eta = E\eta_1 \times \cdots \times E\eta_m$.

Proof. If $|E\eta_i| < \infty$ then by the definition of the Lebesgue integral we have $E|\eta_i| < \infty$, $1 \le i \le m$. Let

$$\chi_+(t) = \begin{cases} 1, & t \ge 0 \\ 0, & t < 0 \end{cases}$$

$$\chi_-(t) = \begin{cases} 0, & t \ge 0 \\ 1, & t < 0 \end{cases}$$

Then $\chi_+ + \chi_- = 1$, $\eta_i = \eta_i(\chi_+(\eta_i) + \chi_-(\eta_i)) = \eta_i \chi_+(\eta_i) + \eta_i \chi_-(\eta_i) = \eta_{i,+} - \eta_{i,-}$, where $\eta_{i,+} \ge 0$ and $\eta_{i,-} \ge 0$. Substituting these expressions into η and multiplying out we obtain η as a linear combination of products of the form $\eta_{1,\epsilon_1} \times \cdots \times \eta_{m,\epsilon_m}$, where the ϵ_i take the values $+$ or $-$. In each such product the $\eta_{1,\epsilon_1}, \ldots \eta_{m,\epsilon_m}$ form a collection of jointly independent random variables as follows from Lemma 12.1. Therefore it suffices to prove the lemma for non-negative random variables.

So let η_i, $1 \le i \le m$ be non-negative.

We introduce the functions $g_i^{(n)}(x)$, where $g_i^{(n)}(x) = k2^{-n}$ if $k2^{-n} \le g_i(x) < (k+1)2^{-n}$. Then $g_i^{(n)}(x) \uparrow g_i(x)$ as $n \to \infty$. Let us form the random variables $\eta_i^{(n)} = g_i^{(n)}(\xi_i)$, $1 \le i \le m$. By Lemma 12.1 they are jointly independent. In addition $\eta_i^{(n)} \uparrow \eta_i$, and $\eta^{(n)} = \prod_{i=1}^{m} \eta_i^{(n)} \uparrow \eta$ as $n \to \infty$. By definition of the Lebesgue integral we have

$$E\eta = \lim_{n \to \infty} E\eta^{(n)}.$$

Furthermore

$$E\eta^{(n)} = \sum_{k_1,\ldots,k_m \ge 0} \prod_{i=1}^{m} k_i 2^{-n} P(\{\omega | \eta_i^{(n)} = k_i 2^{-n}, 1 \le i \le n\})$$

$$= \sum_{k_1,\ldots,k_m \ge 0} \prod_{i=1}^{m} k_i 2^{-n} \prod_{i=1}^{m} P(\{\omega | \eta_i^{(n)} = k_i 2^{-n}\})$$

by independence of the $\eta_i^{(n)}$. The latter expression is a series with non-negative terms. We interchange the order of summation and multiplication:

$$\sum_{k_1,\ldots,k_m \ge 0} \prod_{i=1}^{m} k_i 2^{-n} \prod_{i=1}^{m} P(\{\omega | \eta_i^{(n)} = k_i 2^{-n}\})$$

$$= \prod_{i=1}^{m} \sum_{k_i} k_i 2^{-n} P(\{\omega | \eta_i^{(n)} = k_i 2^{-n}\}) = \prod_{i=1}^{m} E\eta_i^{(n)} \xrightarrow[n \to \infty]{} \prod_{i=1}^{m} E\eta_i.$$

Thus $E\eta = \lim_{n \to \infty} E\eta^{(n)} = \prod_{i=1}^{m} E\eta_i$. $\qquad\square$

The following lemma will be necessary later for the proof of Kolmogorov's inequality.

Lemma 12.3. *Let $\xi_1, \ldots, \xi_n$ be real-valued jointly independent random variables, and let $g(x_1, \ldots, x_k)$ be a Borel function of k variables. Then $\eta_1 = g(\xi_1, \ldots, \xi_k)$, $\xi_{k+1}, \ldots, \xi_n$ form a collection of jointly independent random variables.*

Proof. Let $A, A_{k+1}, \ldots, A_n$ be Borel subsets of the line. We need to prove that $P(\eta_1 \in A, \xi_{k+1} \in A_{k+1}, \ldots, \xi_n \in A_n) = P(\eta_1 \in A) \prod_{i=k+1}^{n} P(\xi_i \in A_i)$. We introduce $A' = \{(x_1, \ldots, x_k) | \eta_1 = g(x_1, \ldots, x_k) \in A\}$. Then A' is a Borel subset of $\mathbb{R}^k$ and we need to prove that

$$P(\eta_1 \in A, \xi_{k+1} \in A_{k+1}, \ldots, \xi_n \in A_n) = P((\xi_1, \ldots, \xi_k) \in A',$$

$$\xi_{k+1} \in A_{k+1}, \ldots, \xi_n \in A_n) = P((\xi_1, \ldots, \xi_k) \in A') \prod_{i=k+1}^{n} P(\xi_i \in A_i).$$

If A' is a rectangle, i.e. if $A' = A'_1 \times A'_2 \times \cdots \times A'_k$ then the desired equality follows from the independence of $\xi_1, \ldots, \xi_n$. If A' is a finite or countable union of disjoint rectangles then the desired equality follows from the σ-additivity of probability measures. As is shown in general measure theory, any Borel subset of $\mathbb{R}^k$ can be approximated by such unions at any arbitrary precision, where precision is defined by the measure of the symmetric difference. $\quad\square$

We now turn to one of the fundamental theorems of probability theory, the Law of Large Numbers. First we introduce the following general definition.

Definition 12.2. Let $\{\varsigma_1, \ldots, \varsigma_n, \ldots\}$, where $\varsigma_i = f_i(\omega)$, be a sequence of real-valued random variables. We say that this sequence converges to $\varsigma = f(\omega)$:

- in probability (in measure) if $\lim_{n \to \infty} P(|\varsigma_n - \varsigma| > \epsilon) = 0$ for any $\epsilon > 0$;
- almost surely if $\lim_{n \to \infty} f_n(\omega) = f(\omega)$ for almost all ω.

It is clear that almost sure convergence implies convergence in probability but the converse is false, as simple examples show.

Let $\xi_1, \xi_2, \ldots, \xi_n, \ldots$ be a sequence of random variables with finite mathematical expectations $m_i = E\xi_i$, $i = 1, 2, \ldots$. We form the arithmetic mean $\varsigma_n = \frac{1}{n}(\xi_1 + \cdots + \xi_n)$ and $\bar{m}_n = \frac{1}{n}(m_1 + \cdots + m_n)$.

Definition 12.3. The sequence of random variables $\{\xi_i\}$ satisfies

- the Law of Large Numbers if $\varsigma_n - \bar{m}_n$ converges in probability to 0, i.e. $P(|\varsigma_n - \bar{m}_n| > \epsilon) \to 0$ as $n \to \infty$ for any $\epsilon > 0$;
- the Strong Law of Large Numbers if $\varsigma_n - \bar{m}_n$ converges to 0 almost surely, i.e. $\lim_{n \to \infty}(\varsigma_n - \bar{m}_n) = 0$ for almost all ω.

If the random variables ξ_i are jointly independent and if $\operatorname{Var}\xi_i \le V = $ const, then it follows by Chebyshev's Inequality that the law of large numbers holds:

$$P(|\varsigma_n - \bar{m}_n| > \epsilon) = P\big(|(\xi_1 + \cdots + \xi_n) - (m_1 + \cdots + m_n)| \ge \epsilon n\big)$$
$$\le \operatorname{Var}(\xi_1 + \cdots + \xi_n)/\epsilon^2 n^2.$$

In the case of independent random variables $\operatorname{Var}(\xi_1 + \cdots + \xi_n) = \operatorname{Var}\xi_1 + \cdots + \operatorname{Var}\xi_n \le Vn$ (see Lecture 4). Therefore the last expression is less or equal to $Vn/\epsilon^2 n^2 = V/\epsilon^2 n \to 0$ as $n \to \infty$. We also have a stronger theorem due to Khinchin:

Theorem 12.1 (Khinchin). *A sequence ξ_i of jointly independent identically distributed random variables with finite mathematical expectation satisfies the Law of Large Numbers.*

Historically Khinchin's Theorem was one of the first theorems connected with the Law of Large Numbers. We will not prove it, but we will derive it from an analogous theorem of Kolmogorov connected with the Strong Law of Large Numbers, to which we now turn. We need two sufficiently general statements.

Lemma 12.4 (First Borel-Cantelli). *Let $\{A_n\}$ be a sequence of events on a probability space $(\Omega, \mathcal{F}, P)$ with $\sum P(A_n) < \infty$. If $A = \{\omega|$ there exists an infinite sequence $n_i = n_i(\omega)$ for which $\omega \in A_{n_i}, i = 1, 2, \ldots\}$ then $P(A) = 0$.*

Proof. We write A in the following form:

$$A = \bigcap_{k=1}^{\infty} \bigcup_{n=k}^{\infty} A_n.$$

Then $P(A) \le P(\bigcup_{n=k}^{\infty} A_n) \le \sum_{n=k}^{\infty} P(A_n) \to 0$ as $k \to \infty$. $\square$

Lemma 12.5 (Second Borel-Cantelli). *Let $\{A_n\}$ be a sequence of jointly independent events on a probability space with $\sum P(A_n) = \infty$, and let $A = \{\omega|$ there exists an infinite sequence $n_i(\omega)$ such that $\omega \in A_{n_i}\}$; then $P(A) = 1$.*

Proof. We have $\bar{A} = \bigcup_{k=1}^{\infty} \bigcap_{n=k}^{\infty} \bar{A}_n$. Then $P(\bar{A}) \le \sum_{k=1}^{\infty} P(\bigcap_{k=n}^{\infty} \bar{A}_n)$ for any n. By independence of the A_n we have the independence of the $\bar{A}_n$ and therefore $P(\bigcap_{n=k}^{\infty} \bar{A}_n) = \prod_{n=k}^{\infty}(1 - P(A_n)) = 0$ since $\sum_{n=1}^{\infty} P(A_n) = \infty$. $\square$

Let $\xi_1, \xi_2, \ldots$ be a sequence of jointly independent random variables which have a finite mathematical expectation and variance, i.e. $m_i = E\xi_i$, $V_i = \operatorname{Var}\xi_i < \infty$, $1 \le i \le n$. Then

$$\operatorname{Var}(\xi_1 + \cdots + \xi_n) = \operatorname{Var}\xi_1 + \cdots + \operatorname{Var}\xi_n = \sum_{i=1}^{n} V_i.$$

Theorem 12.2 (Kolmogorov's Inequality).

$$P\left\{\max_{1\le k\le n} |(\xi_1 + \cdots + \xi_k) - (m_1 + \cdots + m_k)| \ge t\right\} \le \frac{1}{t^2}\sum_{i=1}^{n} V_i.$$

Proof. Let us introduce the events $C_k = \{\omega| \, |(\xi_1 + \cdots + \xi_i) - (m_1 + \cdots + m_i)| < t$ for $1 \le i < k, |(\xi_1 + \cdots + \xi_k) - (m_1 + \cdots + m_k)| \ge t\}$, $C = \bigcup_{k=1}^{n} C_k$. It is clear that C is the event whose probability appears in Kolmogorov's inequality, and that the C_k do not intersect. We have

$$\mathrm{Var}(\xi_1 + \cdots + \xi_n) = \int_\Omega \left((\xi_1 + \cdots + \xi_n) - (m_1 + \cdots + m_n)\right)^2 dP$$

$$\ge \int_C \left((\xi_1 + \cdots + \xi_n) - (m_1 + \cdots + m_n)\right)^2 dP$$

$$= \sum_{k=1}^{n} \int_{C_k} \left((\xi_1 + \cdots + \xi_n) - (m_1 + \cdots + m_n)\right)^2 dP$$

$$= \sum_{k=1}^{n} \left[\int_{C_k} \left((\xi_1 + \cdots + \xi_k) - (m_1 + \cdots + m_k)\right)^2 dP\right.$$

$$+ 2\int_{C_k} \left((\xi_1 + \cdots + \xi_k) - (m_1 + \cdots + m_k)\right)$$

$$\times \left((\xi_{k+1} + \cdots + \xi_n) - (m_{k+1} + \cdots + m_n)\right) dP$$

$$\left. + \int_{C_k} \left((\xi_{k+1} + \cdots + \xi_n) - (m_{k+1} + \cdots + m_n)\right)^2 dP\right].$$

The last integral is non-negative. The main point is that the middle integral is equal to zero. Indeed set

$$g(x_1,\ldots,x_k) = \begin{cases} 1 & \text{if } |(x_1 + \cdots + x_i) - (m_1 + \cdots + m_i)| < t \text{ for } 1 \le i < k, \\ & \text{and } |(x_1 + \cdots + x_k) - (m_1 + \cdots + m_k)| \ge t, \\ 0 & \text{otherwise.} \end{cases}$$

Then $g(\xi_1,\ldots,\xi_k)$ is the indicator function of C_k. Therefore

$$\int_{C_k} \left((\xi_1 + \cdots + \xi_k) - (m_1 + \cdots + m_k)\right)$$

$$\times \left((\xi_{k+1} + \cdots + \xi_n) - (m_{k+1} + \cdots + m_n)\right) dP$$

$$= \sum_{i=k+1}^{n} \int_\Omega g(\xi_1,\ldots,\xi_k)\left((\xi_1 + \cdots + \xi_k) - (m_1 + \cdots + m_k)\right)(\xi_i - m_i)\, dP.$$

Applying Lemma 12.3 to $g_1(x_1,\ldots,x_k) = g(x_1,\ldots,x_k)\left((x_1 + \cdots + x_k) - (m_1 + \cdots + m_k)\right)$ we find that the random variables $(\xi_i - m_i)$ and $g_1(\xi_1,\ldots,\xi_k)$ are independent for $i > k$. Therefore by Lemma 12.1 we have

$$\int_\Omega g(\xi_1,\ldots,\xi_k)\big((\xi_1+\cdots+\xi_k)-(m_1+\cdots+m_k)\big)(\xi_i-m_i)\,dP$$

$$=\int_\Omega g_1(\xi_1,\ldots,\xi_k)(\xi_i-m_i)\,dP$$

$$=\int_\Omega g_1(\xi_1,\ldots,\xi_k)\,dP\int_\Omega(\xi_i-m_i)\,dP=0.$$

Finally

$$\sum_{i=1}^n V_i \geq \sum_{k=1}^n \int_{C_k}\big((\xi_1+\cdots+\xi_k)-(m_1+\cdots+m_k)\big)^2\,dP$$

$$\geq t^2\sum_{k=1}^n P(C_k)=t^2 P(C),$$

i.e.

$$P(C)\leq\frac{1}{t^2}\sum_{i=1}^n V_i.$$

□

Theorem 12.3 (First Kolmogorov). *A sequence of jointly independent random variables* $\{\xi_i\}$, *such that* $\sum_{i=1}^\infty \frac{1}{i^2}\mathrm{Var}\,\xi_i<\infty$ *satisfies the Strong Law of Large Numbers.*

Proof. Let $\xi_i'=\xi_i-m_i$, $\varsigma_n'=\frac{1}{n}\big((\xi_1+\cdots+\xi_n)-(m_1+\cdots+m_n)\big)=\frac{1}{n}\sum_{i=1}^n \xi_i'$. We need to show that $\varsigma_n'\to 0$ almost everywhere. Let us choose $\epsilon>0$ and form the event $B(\epsilon)=\{\omega|$ there exists $N=N(\omega)$, such that for all $n\geq N(\omega)$ we have $|\varsigma_n'|<\epsilon\}$. We have

$$B(\epsilon)=\bigcup_{N=1}^\infty\bigcap_{n=N}^\infty\{\omega|\,|\varsigma_n'|<\epsilon\}.$$

We define

$$B_m(\epsilon)=\{\omega|\max_{2^{m-1}\leq n<2^m}|\varsigma_n'|\geq\epsilon\}.$$

By Kolmogorov's inequality we have

$$P\{B_m(\epsilon)\} = P\Big(\max_{2^{m-1} \leq n < 2^m} \frac{1}{n}\Big|\sum_{i=1}^{n} \xi_i'\Big| \geq \epsilon\Big)$$

$$= P\Big(\max_{2^{m-1} \leq n < 2^m} \Big|\sum_{i=1}^{n} \xi_i'\Big| \geq \epsilon n\Big)$$

$$\leq P\Big(\max_{2^{m-1} \leq n < 2^m} \Big|\sum_{i=1}^{n} \xi_i'\Big| \geq \epsilon 2^{m-1}\Big)$$

$$\leq P\Big(\max_{1 \leq n \leq 2^m} \Big|\sum_{i=1}^{n} \xi_i'\Big| \geq \epsilon 2^{m-1}\Big)$$

$$\leq \frac{1}{\epsilon^2 2^{2m-2}} \sum_{i=1}^{2^m} \mathrm{Var}\,\xi_i.$$

Therefore

$$\sum_{m=1}^{\infty} P\{B_m(\epsilon)\} \leq \sum_{m=1}^{\infty} \frac{1}{\epsilon^2 2^{2m-2}} \sum_{i=1}^{2^m} \mathrm{Var}\,\xi_i$$

$$= \frac{1}{\epsilon^2} \sum_{i=1}^{\infty} \mathrm{Var}\,\xi_i \sum_{\substack{m' \geq m_i \\ 2^{m_i-1} \leq i < 2^{m_i}}} \frac{1}{2^{2m'-2}}$$

$$\leq \frac{4}{\epsilon^2} \sum_{i=1}^{\infty} \mathrm{Var}\,\xi_i \frac{4}{2^{2m_i}3} \leq \frac{16}{\epsilon^2} \sum_{i=1}^{\infty} \frac{\mathrm{Var}\,\xi_i}{i^2} < \infty$$

by assumption. By the Borel-Cantelli Lemma we know that for almost all ω there exists an integer $m_0 = m_0(\omega)$ such that $\max_{2^{m-1} \leq n < 2^m} |\varsigma_n'| < \epsilon$ for all $m \geq m_0$. Therefore $P(B(\epsilon)) = 1$ for any $\epsilon > 0$. In particular $P(B(1/k)) = 1$ and $P(\bigcap_k B(1/k)) = 1$. But if $\omega \in \bigcap_{k=1}^{\infty} B(1/k)$ then, for any k, $\omega \in B(1/k)$, i.e. there exists $N = N(\omega, k)$ such that for all $n \geq N(\omega, k)$ we have $|\varsigma_n'| < 1/k$. In other words $\lim_{n \to \infty} \varsigma_n' = 0$ for such ω.   □

Theorem 12.4 (Second Kolmogorov). *A sequence $\{\xi_i\}$ of jointly independent identically distributed random variables with finite mathematical expectation $m = E\xi_i$ satisfies the Strong Law of Large Numbers.*

Proof. By introducing $\xi_i' = \xi_i - m_i$ we reduce the problem to the case where $m = 0$. Let us set

$$\eta_i(\omega) = \begin{cases} \xi_i(\omega) & \text{if } |\xi_i| \leq i, \\ 0 & \text{otherwise.} \end{cases}$$

By the properties of the Lebesgue integral we then have

$$m_i = E\eta_i = \int_{|x| \leq i} x\, dF_\xi(x) \to \int x\, dF_\xi(x) = 0 \quad \text{as} \quad i \to \infty,$$

where F_ξ is the common distribution function of the random variables ξ_i. We have

$$\frac{1}{n}(\xi_1 + \cdots + \xi_n) = \frac{1}{n}(\eta_1 + \cdots + \eta_n) - \frac{1}{n}(m_1 + \cdots + m_n)$$

$$+ \frac{1}{n}\sum_{i=1}^{n}(\xi_i - \eta_i) + \frac{1}{n}(m_1 + \cdots + m_n).$$

Since $m_n \to 0$ as $n \to \infty$ the latter arithmetic mean converges to 0 as $n \to \infty$.

We now show that with probability 1, $\xi_i \neq \eta_i$ only for a finite number of values of i, which implies the convergence to zero, almost everywhere, of the term $\frac{1}{n}\sum_{i=1}^{n}(\xi_i - \eta_i)$. For this we once again use the Borel-Cantelli Lemma:

$$\begin{aligned}
\sum_{i=1}^{\infty} P\{\xi_i \neq \eta_i\} &= \sum_{i=1}^{\infty} P\{|\xi_i| \geq i\} \\
&= \sum_{i=1}^{\infty}\sum_{k=i}^{\infty} P\{k \leq |\xi_i| < k+1\} \\
&= \sum_{k=0}^{\infty}(k+1)P\{k \leq |\xi_i| < k+1\} \\
&= \sum_{k=0}^{\infty} kP\{k \leq |\xi_i| < k+1\} + 1.
\end{aligned} \tag{12.1}$$

The last sum is the mathematical expectation of the random variable η where

$$\eta = \{k, \text{ if } k \leq |\xi_1(\omega)| < k+1\}.$$

It is clear that $\eta \leq |\xi_1|$. Therefore by the properties of the Lebesgue integral we have $E(|\xi_1| - \eta) \geq 0$ and $E|\xi_1| - E\eta \geq 0$, i.e. $E|\xi_1| \geq E\eta$. We recall that the inequality $|E\xi| < \infty$ always means that $E|\xi| < \infty$. Thus the series in (12.1) converges.

We show that the First Kolmogorov Theorem applies to the first sum. We have

$$\mathrm{Var}\eta_i = E(\eta_i - m_i)^2 \leq E\eta_i^2 = \int_{|x| \leq i} x^2 \, dF_\xi(x)$$

$$\leq \sum_{i=1}^{\infty} \frac{1}{i^2} \int_{|x| \leq i} x^2 \, dF_\xi(x) = \sum_{i=1}^{\infty} \frac{1}{i^2} \sum_{k=1}^{i} \int_{k+1 \leq |x| < k} x^2 \, dF_\xi(x)$$

$$= \sum_{k=1}^{\infty} \int_{k-1 \leq |x| < k} x^2 \, dF_\xi(x) \sum_{i=k}^{\infty} \frac{1}{i^2}.$$

We use the fact that for $k > 1$, $\sum_{i=k}^{\infty} \frac{1}{i^2} \leq \sum_{i=k}^{\infty} \frac{1}{i(i-1)} = \sum_{i=k}^{\infty}\left(\frac{1}{i-1} - \frac{1}{i}\right) = \frac{1}{k-1}$. So the latter series is no greater than

$$\int_{|x|\leq 1} x^2 \, dF_\xi(x) \sum_{i=1}^{\infty} \frac{1}{i^2} + \sum_{k=2}^{\infty} \frac{1}{k-1} \int_{k-1\leq |x|\leq k} x^2 \, dF_\xi(x)$$

$$= \text{const} \; + \text{const} \sum_{k=1}^{\infty} \int_{k-1\leq |x|\leq k} |x| \, dF_\xi(x)$$

$$\leq \text{const} \; E|\xi| < \infty.$$

$\square$

The Law of Large Numbers, as well as the Strong Law of Large Numbers, are related to theorems known as the Ergodic theorems. These theorems provide general conditions under which the arithmetic means of random variables have a limit. Basically these conditions reduce to stationarity, i.e. the assumption that for a sequence of random variables $\{\xi_n\}$ the distribution of any collection of k random variables $\xi_{n+1}, \ldots, \xi_{n+k}$ does not depend on n.

The statements of both laws of large numbers are that for a sequence of random variables $\{\xi_n\}$ the arithmetic mean $\frac{1}{n} \sum_{k=1}^{n} \xi_k$ is close to its mathematical expectation and therefore asymptotically does not depend on ω, i.e. is not random. In other words deterministic regularities appear with high probabilities in long series of random variables. In daily life we constantly encounter the Law of Large Numbers. For example the fact that in its equilibrium state a gas occupies all the available volume is a statement of a type of law of large numbers.

Lecture 13. Weak Convergence of Probability Measures on the Line and Helly's Theorems

In the proof of the De Moivre-Laplace Limit Theorem (see Lecture 3) we came across the following situation. We considered a random variable η_n which takes values of the form $k/\sqrt{npq}$ where $k > 0$ is an integer, with probabilities of the form

$$\frac{1}{\sqrt{2\pi npq}} \exp\left(-\frac{1}{2}\frac{k^2}{npq}\right)\left(1 + O\left(\frac{1}{\sqrt{n}}\right)\right).$$

We do not specify the details of this expression here. The main point is that the random variable η_n has a discrete distribution which converges in a natural sense to a continuous distribution with Gaussian density $(1/\sqrt{2\pi})\exp(-x^2/2)$ as $n \to \infty$.

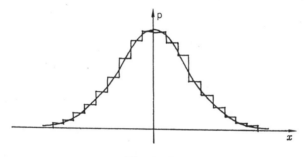

Fig. 13.1.

This form of convergence of distributions is fundamental to limit theorems in probability theory, which we now consider. We now examine more explicitly several concepts which are related to this form of convergence.

Let X be a metric space, $\mathcal{B}(X)$ be the σ-algebra of its Borel subsets, and P_n be a sequence of probability measures on $(X, \mathcal{B}(X))$.

Definition 13.1. The sequence $\{P_n\}$ converges weakly to the probability measure P if, for any bounded continuous function $f \in C(X)$,

$$\lim_{n \to \infty} \int f(x)\, dP_n(x) = \int f(x)\, dP(x).$$

This form of convergence is sometimes written as: $P_n \Rightarrow P$. Let us consider a sequence of real-valued random variables $\{\xi_n\}$ with corresponding probability measures on the line $\{P_{\xi_n}\}$, i.e. $P_{\xi_n}(A) = P(\{\omega | \xi_n \in A\})$. Then $\{\xi_n\}$ converges in distribution if the sequence $\{P_{\xi_n}\}$ converges weakly to P, i.e. $P_{\xi_n} \Rightarrow P$.

In Definition 13.1 we could omit the requirement that P_n and P be probability measures, i.e. normed measures on X. We then obtain a definition of weak convergence for arbitrary finite measures given on $\mathcal{B}(X)$. This remark will be useful below.

To each random variable ξ corresponds a distribution function $F_\xi(x)$ which uniquely determines the distribution P_ξ. We now express the condition of weak convergence in terms of distribution functions.

Let $P_n \Rightarrow P$ and $\{F_n\}$ and F be the corresponding distribution functions, i.e. $F_n(x) = P_n((-\infty, x])$, $F(x) = P((-\infty, x])$. Recall that x is a continuity point of $F(x)$ if and only if $P(\{x\}) = 0$.

Theorem 13.1. *The sequence $P_n \Rightarrow P$ if and only if $F_n(x) \to F(x)$ for every continuity point x of the function F.*

Proof. Let $P_n \Rightarrow P$ and let x be a continuity point of F. We consider the functions f, f_δ^+ and f_δ^- whose graphs are given in Fig. 13.2.

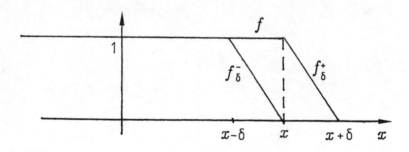

Fig. 13.2.

Formally we have

$$f(y) = \begin{cases} 1, & y \le x \\ 0, & y > x \end{cases}$$

$$f_\delta^+(y) = \begin{cases} 1, & y \le x \\ 1 - (y - x)/\delta, & x < y \le x + \delta \\ 0, & y > x + \delta \end{cases}$$

$$f_\delta^-(y) = \begin{cases} 1, & y \le x - \delta \\ 1 - (y - x - \delta)/\delta, & x - \delta < y \le x \\ 0, & y > x \end{cases}$$

The functions f_δ^+ and f_δ^- are continuous and $f_\delta^- < f < f_\delta^+$. Using the fact that x is a continuity point of F we have, for any $\epsilon > 0$ and $n \ge n_0(\epsilon)$,

$$F_n(x) = \int f(y)\, dF_n(y) \le \int f_\delta^+(y)\, dF_n(y)$$

$$\le \int f_\delta^+(y)\, dF(y) + \frac{\epsilon}{2} \le F(x+\delta) + \frac{\epsilon}{2} \le F(x) + \epsilon,$$

if δ is such that $|F(x \pm \delta) - F(x)| \le \epsilon/2$. On the other hand, for such n we have

$$F_n(x) = \int f(y)\, dF_n(y) \ge \int f_\delta^-(y)\, dF_n(y)$$

$$\ge \int f_\delta^-(y)\, dF(y) - \epsilon/2 \ge F(x-\delta) - \epsilon/2 \ge F(x) - \epsilon.$$

In other words, $|F_n(x) - F(x)| \le \epsilon$ for all sufficiently large n.

We now prove the converse. Let $F_n(x) \to F(x)$ at every continuity point of F. We prove that

$$\int f(x)\, dF_n(x) \to \int f(x)\, dF(x)$$

for any bounded function f.

Let $M = \sup |f(x)|$. The discontinuity points of the function F are those isolated points x_i whose probability is positive (the probability is equal to the height of the jump in the graph of F at the point x_i). It follows that the set of discontinuity points is at most countable.

Let us fix $\epsilon > 0$. We choose $K_1(\epsilon) = K_1$, $K_2(\epsilon) = K_2$, $K_1 < K_2$ such that

a_1) K_1, K_2 are continuity points of F;
a_2) $P\{(K_2, \infty)\} = 1 - F(K_2) \le \epsilon/16M$;
a_3) $P\{(-\infty, K_1]\} = F(K_1) \le \epsilon/16M$.

Since K_1 and K_2 are continuity points of F, for $n \ge n_1(\epsilon)$ we have:

$$1 - F_n(K_2) \le \epsilon/8M, \quad F_n(K_1) \le \epsilon/8M.$$

Therefore

$$I_n = \int f\, dF_n - \int f\, dF$$

$$= \int_{(K_1, K_2]} f\, dF_n - \int_{(K_1, K_2]} f\, dF + \int_{(-\infty, K_1]} f\, dF_n + \int_{(K_2, \infty)} f\, dF_n$$

$$- \int_{(-\infty, K_1]} f\, dF - \int_{(K_2, \infty)} f\, dF.$$

Each of the four latter integrals can be estimated in a simple way: for $n \ge n_1(\epsilon)$ we have

$$\left| \int_{(-\infty, K_1]} f \, dF_n \right| \le \int_{(-\infty, K_1]} |f| \, dF_n \le M F_n(K_1) \le \epsilon/8;$$

$$\left| \int_{(K_2, \infty)} f \, dF_n \right| \le \int_{(K_2, \infty)} |f| \, dF_n \le M(1 - F_n(K_2)) \le \epsilon/8;$$

$$\left| \int_{(-\infty, K_1]} f \, dF \right| \le \epsilon/16; \quad \left| \int_{(K_2, \infty)} f \, dF \right| \le \epsilon/16.$$

Therefore

$$\left| \int_{(-\infty, K_1]} f \, dF_n + \int_{(K_2, \infty)} f \, dF_n - \int_{(-\infty, K_1]} f \, dF - \int_{(K_2, \infty)} f \, dF \right|$$

$$\le \left| \int_{(-\infty, K_1]} f \, dF_n \right| + \left| \int_{(K_2, \infty)} f \, dF_n \right| + \left| \int_{(-\infty, K_1]} f \, dF \right|$$

$$+ \left| \int_{(K_2, \infty)} f \, dF \right| \le 3\epsilon/8.$$

It now remains to estimate the difference $I'_n = \int_{(K_1, K_2]} f \, dF_n - \int_{(K_1, K_2]} f \, dF$. We choose a collection $x_0 = K_1 < x_1 < \ldots < x_p = K_2$ of continuity points of the function F such that $x_{i+1} - x_i \le \delta$, where δ is chosen so that $|f(x') - f(x'')| \le \epsilon/4$, if $K_1 \le x', x'' \le K_2$, $|x' - x''| \le \delta$. We define a new function $f^{(\delta)}$ on $(K_1, K_2]$ by

$$f^{(\delta)}(x) = f^{(\delta)}(x_i) \quad \text{if} \quad x_i < x \le x_{i+1}, \quad i = 0, \ldots, p - 1.$$

Then $|f^{(\delta)}(x) - f(x)| \le \epsilon/4$ and

$$\left| \int_{(K_1, K_2]} f \, dF_n - \int_{(K_1, K_2]} f^{(\delta)} \, dF_n \right| \le \int_{(K_1, K_2]} |f - f^{(\delta)}| \, dF_n \le \epsilon/4,$$

$$\left| \int_{(K_1, K_2]} f \, dF - \int_{(K_1, K_2]} f^{(\delta)} \, dF \right| \le \int_{(K_1, K_2]} |f - f^{(\delta)}| \, dF \le \epsilon/4.$$

Furthermore

$$|I'_n| = \left| \int_{(K_1, K_2]} f \, dF_n - \int_{(K_1, K_2]} f \, dF \right|$$

$$\le \left| \int_{(K_1, K_2]} f \, dF_n - \int_{(K_1, K_2]} f^{(\delta)} \, dF_n \right|$$

$$+ \left| \int_{(K_1, K_2]} f \, dF - \int_{(K_1, K_2]} f^{(\delta)} \, dF \right|$$

$$+ \left| \int_{(K_1, K_2]} f^{(\delta)} \, dF_n - \int_{(K_1, K_2]} f^{(\delta)} \, dF \right|$$

$$\le \frac{\epsilon}{2} + \left| \int_{(K_1, K_2]} f^{(\delta)} \, dF_n - \int_{(K_1, K_2]} f^{(\delta)} \, dF \right|.$$

We now note that from the definition

$$\int_{(K_1, K_2]} f^{(\delta)} \, dF_n = \sum_{i=0}^{p-1} f(x_i)(F_n(x_{i+1}) - F_n(x_i)),$$

$$\int_{(K_1, K_2]} f^{(\delta)} \, dF = \sum_{i=0}^{p-1} f(x_i)(F(x_{i+1}) - F(x_i))$$

and

$$\left| \int_{(K_1, K_2]} f^{(\delta)} \, dF_n - \int_{(K_1, K_2]} f^{(\delta)} \, dF \right| = \left| \sum_{i=0}^{p-1} f(x_i)(F_n(x_{i+1}) - F_n(x_i)) \right.$$

$$\left. - \sum_{i=0}^{p-1} f(x_i)(F(x_{i+1}) - F(x_i)) \right|$$

$$\leq \sum_{i=0}^{p-1} |f(x_i)| \big(|F_n(x_{i+1}) - F(x_{i+1})| + |F_n(x_i) - F(x_i)| \big).$$

We choose $n_2(\epsilon)$ to be large enough so that $|F_n(x_i) - F(x_i)| \leq \epsilon/16Mp$. This is possible because the points x_i are continuity points of F. The latter expression is then at most $Mp2(\epsilon/16Mp) = \epsilon/8$. Putting together all the estimates we find that $|I_n| \leq \epsilon$ for $n \geq \max(n_1(\epsilon), n_2(\epsilon))$. $\qquad \square$

Definition 13.2. A family of probability measures $\{P_\alpha\}$ on a metric space X is said to be weakly compact if from any subsequence P_{α_n}, $n = 1, 2, \ldots$, one can extract a weakly convergent subsequence $P_{\alpha_{n_k}}$, $k = 1, 2, \ldots$, i.e. $P_{\alpha_{n_k}} \Rightarrow P$. Note that it is not required that $P \in \{P_\alpha\}$.

In what follows we need the following general fact from functional analysis.

Theorem 13.2. *Let X be a separable compact metric space, i.e. a complete separable metric space which is a compact space. Then any family of measures μ_α on $(X, \mathcal{B}(X))$ such that $\mu_\alpha(X) \leq$ const, where const does not depend on α, is weakly compact.*

With the help of this theorem we derive the important Helly's Theorem:

Theorem 13.3 (Helly). *A family of probability measures $\{P_\alpha\}$ on the line $X = \mathbb{R}^1$ is weakly compact if and only if for any $\epsilon > 0$ there exists an interval $[-K, K]$ for which $P_\alpha(\mathbb{R}^1 \setminus [-K, K]) \leq \epsilon$ for all α.*

Proof. Assuming that the latter inequality is satisfied, we first prove the weak compactness of the family $\{P_\alpha\}$. We take a sequence of probability measures from the family $\{P_\alpha\}$ which, for simplicity, we write in the form P_n, $n = 1, 2, \ldots$, and let $X = [-\ell, \ell]$. For any integer ℓ, X is a separable, compact metric space, and $P_n([-\ell, \ell]) \leq 1$ for all n. Therefore for any ℓ we can use the general theorem from functional analysis that was quoted above.

Considering the interval $[-1,1]$ we find a subsequence $P_{n_i^{(1)}}$, $i = 1, 2, \ldots$, such that the restrictions of all the $P_{n_i^{(1)}}$ to $[-1,1]$ converge weakly to a limit which we denote by $P^{(1)}$. We emphasize that $P^{(1)}$ is a measure on $[-1,1]$, which is not necessarily normed. We choose a subsequence $\{n_i^{(2)}\} \subseteq \{n_i^{(1)}\}$ such that the restrictions of $P_{n_i^{(2)}}$ to $[-2,2]$ converge weakly to a limit $P^{(2)}$ which is a measure on $[-2,2]$, and so forth. In other words we construct a decreasing collection of subsequences $\{n_i^{(\ell)}\}$, i.e. $\{n_i^{(\ell+1)}\} \subseteq \{n_i^{(\ell)}\}$, such that the restrictions $P_{n_i^{(\ell)}}$ to $[-\ell,\ell]$ converge to a measure on $[-\ell,\ell]$, which we denote by $P^{(\ell)}$. It is essential that these measures be compatible: if $A \subset [-\ell_1,\ell_1] \subset [-\ell_2,\ell_2]$ then $P^{(\ell_1)}(A) = P^{(\ell_2)}(A)$. This holds since $\{n_i^{(\ell+1)}\} \subseteq \{n_i^{(\ell)}\}$ for all ℓ. For any Borel set $A \in \mathcal{B}(\mathbb{R}^1)$ we now set

$$P(A) = P^{(1)}(A \cap [-1,1]) + \sum_{\ell=2}^{\infty} P^{(\ell)}\big(A \cap ([-\ell,\ell] \setminus [-\ell+1,\ell-1])\big).$$

We show that P is a probability measure, and that P is the weak limit of the diagonal subsequence $\{P_{n_i}^{(i)}\}$.

It is clear that $P(A) \geq 0$. Let us check σ-additivity. Let $A = \bigcup_{i=1}^{\infty} A_i$ where the A_i do not intersect pairwise. Then by the σ-additivity of each measure $P^{(\ell)}$

$$
\begin{aligned}
P(A) &= P^{(1)}(A \cap [-1,1]) + \sum_{\ell=2}^{\infty} P^{(\ell)}\big(A \cap ([-\ell,\ell] \setminus [-\ell+1,\ell-1])\big) \\
&= \sum_{i=1}^{\infty} P^{(1)}(A_i \cap [-1,1]) + \sum_{\ell=2}^{\infty} \sum_{i=1}^{\infty} P^{(\ell)}\big(A_i \cap ([-\ell,\ell] \setminus [-\ell+1,\ell-1])\big) \\
&= \sum_{i=1}^{\infty} \big[P^{(1)}(A_i \cap [-1,1]) + \sum_{\ell=2}^{\infty} P^{(\ell)}\big(A_i \cap ([-\ell,\ell] \setminus [-\ell+1,\ell-1])\big)\big] \\
&= \sum_{i=1}^{\infty} P(A_i).
\end{aligned}
$$

It remains to be checked that $P(\mathbb{R}^1) = 1$. It is clear that $P(\mathbb{R}^1) = \lim_{\ell \to \infty} P([-\ell,\ell]) \leq 1$. But by assumption for any $\epsilon > 0$ one can find $\ell(\epsilon)$ such that $P_{n_i^{(1)}}([-\ell,\ell]) \geq 1 - \epsilon$ for all i. Consequently $P\{[-\ell,\ell]\} = P^{(\ell)}\{[-\ell,\ell]\} = \lim_{i \to \infty} P_{n_i^{(\ell)}}\{[-\ell,\ell]\} \geq 1 - \epsilon$. Therefore $P(\mathbb{R}^1) = 1$.

We now show that $P_{n_\ell^{(\ell)}} \Rightarrow P$. We take an arbitrary bounded continuous function f and establish that $\lim_{\ell \to \infty} \int f(x)\, dP_{n_\ell^{(\ell)}}(x) = \int f(x)\, dP(x)$. Let us fix $\epsilon > 0$ and choose a k such that $P(\mathbb{R}^1 \setminus [-k,k]) \leq \epsilon/4M$, where $M = \sup_x |f(x)|$. Then $|\int_{\mathbb{R}^1 \setminus [-k,k]} f(x)\, dP(x)| \leq \epsilon/4$. Furthermore, increasing k if necessary, we can assume that $P_{n_\ell^{(\ell)}}(\mathbb{R}^1 \setminus [-k,k]) \leq \epsilon/4$ for all ℓ. Therefore $|\int_{\mathbb{R}^1 \setminus [-k,k]} f(x)\, dP_{n_\ell^{(\ell)}}(x)| \leq \epsilon/4$. Finally, for $\ell \geq \ell_0(\epsilon)$, we have

$$\left| \int_{[-k,k]} f(x)\, dP_{n_\ell^{(\ell)}}(x) - \int_{[-k,k]} f(x)\, dP(x) \right| \leq \epsilon/2$$

by the weak convergence of the $P_{n_\ell^{(\ell)}}$ on any interval $[-k, k]$. Finally we get
that $| \int f(x)\, dP_{n_\ell^{(\ell)}}(x) - \int f(x)\, dP(x)| \leq \epsilon$ for all sufficiently large ℓ.

We now prove the converse. Let $\{P_\alpha\}$ be a weakly compact family which
does not satisfy the condition in Helly's Theorem. This means that for some
$\epsilon_0 > 0$ and for some integer $k > 0$, there exists α_k such that

$$P_{\alpha_k}([-k,k]) \leq 1 - \epsilon_0.$$

Since $\{P_\alpha\}$ is weakly compact, one can extract from the subsequence
$\{P_{\alpha_k}\}$ a weakly convergent subsequence. Without loss of generality we will
assume that the whole subsequence $P_{\alpha_k} \Rightarrow P$, where P is a probability measure. Let us choose for any k a point β_k, $k - 1 < \beta_k < k$, such that $\pm \beta_k$ are
continuity points of P. Then for all $m \geq k$ we have

$$P_{\alpha_m}((-\beta_k, \beta_k]) \leq P_{\alpha_m}((-k,k]) \leq P_{\alpha_m}((-m,m]) \leq 1 - \epsilon_0.$$

It follows from Theorem 13.1 that

$$\begin{aligned}
P([-\beta_k, \beta_k]) &= F(\beta_k) - F(-\beta_k) \\
&= \lim_{m \to \infty} [F_{\alpha_m}(\beta_k) - F_{\alpha_m}(-\beta_k)] \\
&= \lim_{m \to \infty} P_{\alpha_m}((-\beta_k, \beta_k]) \leq 1 - \epsilon_0.
\end{aligned}$$

Here we have denoted by F and F_{α_m} the distribution functions of the prob-
ability measures P and P_{α_m}. Since $\mathbb{R}^1 = \bigcup_{k=1}^\infty (-\beta_k, \beta_k]$ it follows from the
last inequality that

$$P(\mathbb{R}^1) = \lim_{k \to \infty} P([-\beta_k, \beta_k]) \leq 1 - \epsilon_0,$$

which proves that P is a probability measure. $\square$

Helly's Theorem is a particular case of a much more general theorem due
to Yu. V. Prokhorov.

Theorem 13.4 (Prokhorov). *Let $\{P_\alpha\}$ be a family of probability measures
given on the Borel σ-algebra of a complete separable metric space X. This
family is weakly compact if and only if for any $\epsilon > 0$ there exists a compact
set $K \subset X$ such that $P_\alpha(K) \geq 1 - \alpha$ for all α.*

Lecture 14. Characteristic Functions

In this lecture we consider the Fourier transform for probability measures, and its properties. Let us begin with the case of measures on the line. Let P be a probability measure given on $B(\mathbb{R}^1)$.

Definition 14.1. The characteristic function (Fourier transform) of a measure P is the complex valued function $\phi(\lambda)$ of the variable $\lambda \in \mathbb{R}^1$ given by

$$\phi(\lambda) = \int_{-\infty}^{\infty} e^{i\lambda x} \, dP(x) = \int_{-\infty}^{\infty} \cos \lambda x \, dP(x) + i \int_{-\infty}^{\infty} \sin \lambda x \, dP(x).$$

If $P = P_\xi$ we will denote the characteristic function by $\phi_\xi(\lambda)$ and will call it the characteristic function of the random variable ξ. The expression means that $\phi_\xi(\lambda) = E e^{i\lambda \xi}$. If ξ takes on values $a_1, a_2, \ldots$ with probabilities $p_1, p_2, \ldots$ then

$$\phi_\xi(\lambda) = \sum_{k=1}^{\infty} p_k e^{i\lambda a_k}.$$

If ξ has a probability density $p_\xi(x)$ then

$$\phi_\xi(\lambda) = \int_{-\infty}^{\infty} e^{i\lambda x} p_\xi(x) \, dx.$$

In analysis, the integral on the right hand side is called the Fourier transform of the function $p_\xi(x)$. Therefore in the general case, the notion of the characteristic function of a measure is the generalization of the Fourier transform.

We now establish some simple properties of characteristic functions.

1. $\phi(0) = 1$. This is clear.
2. $|\phi(\lambda)| \leq 1$. Indeed $|\phi(\lambda)| = |\int_{-\infty}^{\infty} e^{i\lambda x} \, dP(x)| \leq \int_{-\infty}^{\infty} dP(x) = 1$.
3. If $\eta = a\xi + b$, where a and b are constants, then

$$\phi_\eta(\lambda) = e^{i\lambda b} \phi_\xi(a\lambda).$$

Indeed $\phi_\eta(\lambda) = E e^{i\lambda \eta} = E e^{i\lambda(a\xi+b)} = E e^{i\lambda b} e^{i\lambda a \xi} = e^{i\lambda b} E e^{i\lambda a \xi} = e^{i\lambda b} \phi_\xi(\lambda a)$.

4. If $\phi_\xi(\lambda_0) = e^{2\pi i a}$ for $\lambda_0 \neq 0$ then ξ is a discrete random variable which takes values of the form $\frac{2\pi}{\lambda_0}(\alpha + m)$, where m is an integer.

Indeed let $\eta = \xi - \frac{2\pi\alpha}{\lambda_0}$. Then from property 3 we have

$$\phi_\eta(\lambda_0) = e^{-2\pi i \alpha} \phi_\xi(\lambda_0) = 1.$$

Furthermore $1 = \phi_\eta(\lambda_0) = E e^{i\lambda_0\eta} = E\cos\lambda_0\eta + iE\sin\lambda_0\eta$. Since $\cos\lambda_0\eta \leq 1$, the latter equality means that $\cos\lambda_0\eta = 1$ with probability 1. This is possible only in the case where η takes values of the form $\eta = \frac{2\pi m}{\lambda_0}$, where m is an integer.

5. $\phi(\lambda)$ is uniformly continuous. Indeed let us take $\epsilon > 0$. We show that there exists a $\delta = \delta(\epsilon)$ such that $|\phi(\lambda) - \phi(\lambda')| < \epsilon$ if $|\lambda - \lambda'| < \delta$.

We first choose t', t'', with $t' < t''$, such that $P\{(t'',\infty)\} \leq \epsilon/6$ and $P\{(-\infty, t')\} \leq \epsilon/6$. Then

$$|\phi(\lambda) - \phi(\lambda')| = \left| \int (e^{i\lambda x} - e^{i\lambda'x}) \, dP(x) \right|$$

$$\leq \left| \int_{x<t'} (e^{i\lambda x} - e^{i\lambda'x}) \, dP(x) \right| + \left| \int_{x>t''} (e^{i\lambda x} - e^{i\lambda'x}) \, dP(x) \right|$$

$$+ \left| \int_{t' \leq x \leq t''} (e^{i\lambda x} - e^{i\lambda'x}) \, dP(x) \right|.$$

Each of the first two terms is no greater in modulus than $\epsilon/3$. Regarding the third term we have

$$\left| \int_{t' \leq x \leq t''} (e^{i\lambda x} - e^{i\lambda'x}) \, dP(x) \right| = \left| \int_{t' \leq x \leq t''} e^{i\lambda x}(1 - e^{i(\lambda'-\lambda)x}) \, dP(x) \right|$$

$$\leq \int_{t' \leq x \leq t''} |1 - e^{i(\lambda'-\lambda)x}| \, dP(x).$$

We set $\delta = \frac{\epsilon}{3\max(|t'|,|t''|)}$. It then follows from the inequality $|1 - e^{i\alpha}| \leq |\alpha|$ that for $|\lambda - \lambda'| \leq \delta$

$$\int_{t' \leq x \leq t''} |1 - e^{i(\lambda-\lambda')x}| \, dP(x) \leq \frac{\epsilon}{3}.$$

Definition 14.2. A complex valued function $f(\lambda)$ is said to be positive semi-definite if, for any $\lambda_1, \ldots, \lambda_r$ the matrix $\|f(\lambda_k - \lambda_\ell)\|$ is positive semi-definite.

This means that the quadratic form $\sum f(\lambda_k - \lambda_\ell)C_k\bar{C}_\ell \geq 0$.

6. Any characteristic function $\phi(\lambda)$ is positive semi-definite. Indeed

$$\sum_{k,\ell=1}^{r} \phi(\lambda_k - \lambda_\ell)C_k\bar{C}_\ell = \sum_{k,\ell=1}^{r} \int e^{i(\lambda_k - \lambda_\ell)}C_k\bar{C}_\ell \, dP(x)$$

$$= \int |\sum_{k=1}^{r} C_k e^{i\lambda_k x}|^2 \, dP(x) \geq 0.$$

By Khinchin's Theorem, any positive semi-definite uniformly continuous function which satisfies the normalization condition $\phi(0) = 1$ is the characteristic function of a probability measure.

7. Assume that the random variable ξ has an absolute moment of order k, i.e. $E|\xi|^k < \infty$. Then $\xi(\lambda)$ is k times continuously differentiable and $\phi_\xi^{(k)}(0) = i^k E\xi^k$.

The proof follows from properties of the Lebesgue integral and consists in checking that the formal differentiation is correct in the equality

$$\phi_\xi^{(k)}(\lambda) = \frac{d^k}{d\lambda^k} \int e^{i\lambda x} dP(x) = i^k \int x^k e^{i\lambda x} dP(x).$$

The last integral is finite, since $E|\xi|^k$ is finite.

In principle all the properties of a probability distribution can be translated using the terminology of characteristic functions, although it often turns out to be a difficult matter. Property 7 shows that the smoothness of the characteristic function is connected with the existence of moments, i.e. with the decrease of P at infinity.

Problem. If the characteristic function $\phi(\lambda)$ is analytic in a neighborhood of $\lambda = 0$, then there exists an $a > 0$ such that $P\{(-\infty, -x)\} \leq \text{const } e^{-ax}$, and $P\{(x, \infty)\} \leq \text{const } e^{-ax}$, for every $x > 0$.

The decrease of $\phi(\lambda)$ at infinity is connected with the existence of a density. For example if $\int_{-\infty}^{\infty} |\phi(\lambda)| \, d\lambda < \infty$ then the distribution P has a density given by

$$p(x) = \frac{1}{2\pi} \int_{-\infty}^{\infty} e^{-i\lambda x} \phi(\lambda) \, d\lambda.$$

We now show that one can always recover the measure P from its characteristic function $\phi(\lambda)$. If $\phi(\lambda)$ is absolutely integrable, then the density of the distribution can be found by the formula given above. Let us assume that P is concentrated at the points $a_1, a_2, \ldots$, and $P\{a_i\} = p_i$. Then $\phi(\lambda) = \sum_s e^{i\lambda a_s} p_s$ and does not tend to 0 as $\lambda \to \infty$. In this case

$$\lim_{R \to \infty} \frac{1}{2R} \int_{-R}^{R} e^{i\lambda x} \phi(\lambda) \, d\lambda = \begin{cases} 0 & \text{for } x \notin (a_1, a_2, \ldots), \\ p_s & \text{for } x = a_s. \end{cases}$$

Before turning to the general case we will say that an interval $[a, b]$ is a continuity interval for P if $P\{a\} = P\{b\} = 0$.

Theorem 14.1. *For any interval* (a, b)

$$\lim_{R \to \infty} \frac{1}{2\pi} \int_{-R}^{R} \frac{e^{-i\lambda a} - e^{-i\lambda b}}{i\lambda} \phi(\lambda) \, d\lambda = P\{(a, b)\} + \frac{1}{2} P\{a\} + \frac{1}{2} P\{b\}.$$

Proof. By Fubini's Theorem on the interchange of the order of integration in a two-dimensional Lebesgue integral, since the integrand is bounded we have

$$\frac{1}{2\pi}\int_{-R}^{R}\frac{e^{-i\lambda a}-e^{-i\lambda b}}{i\lambda}\phi(\lambda)\,d\lambda=\frac{1}{2\pi}\int_{-R}^{R}\frac{e^{-i\lambda a}-e^{-i\lambda b}}{i\lambda}d\lambda\int_{-\infty}^{\infty}e^{i\lambda x}\,dP(x)$$

$$=\frac{1}{2\pi}\int_{-\infty}^{\infty}dP(x)\int_{-R}^{R}\frac{e^{-i\lambda a}-e^{-i\lambda b}}{i\lambda}e^{i\lambda x}\,d\lambda.$$

Furthermore

$$\int_{-R}^{R}\frac{e^{-i\lambda a}-e^{-i\lambda b}}{i\lambda}e^{i\lambda x}\,d\lambda=\int_{-R}^{R}\frac{\cos\lambda(x-a)-\cos\lambda(x-b)}{i\lambda}\,d\lambda$$

$$+\int_{-R}^{R}\frac{\sin\lambda(x-a)-\sin\lambda(x-b)}{\lambda}\,d\lambda.$$

The first integral is equal to 0 since the integrand is odd. In the second integral the integrand is even, therefore

$$\int_{-R}^{R}\frac{e^{-i\lambda a}-e^{-i\lambda b}}{i\lambda}e^{i\lambda x}\,dx=2\int_{0}^{R}\frac{\sin\lambda(x-a)}{\lambda}\,d\lambda-2\int_{0}^{R}\frac{\sin\lambda(x-b)}{\lambda}\,d\lambda.$$

By a change of variables $\mu=\lambda(x-a)$ in the first integral and $\mu=\lambda(x-b)$ in the second integral we obtain

$$2\int_{0}^{R}\frac{\sin\lambda(x-a)}{\lambda}\,d\lambda-2\int_{0}^{R}\frac{\sin\lambda(x-b)}{\lambda}\,d\lambda=2\int_{R(x-b)}^{R(x-a)}\frac{\sin\mu}{\mu}\,d\mu.$$

So

$$\frac{1}{2\pi}\int_{-R}^{R}\frac{e^{-i\lambda a}-e^{-i\lambda b}}{i\lambda}\phi(\lambda)\,d\lambda=\int_{-\infty}^{\infty}dP(x)\frac{1}{\pi}\int_{R(x-b)}^{R(x-a)}\frac{\sin\mu}{\mu}\,d\mu.$$

We now note that for any $t>\pi/2$

$$\left|\int_{0}^{t}\frac{\sin\mu}{\mu}\,d\mu\right|\le\left|\int_{0}^{\pi/2}\frac{\sin\mu}{\mu}\,d\mu\right|+\left|\int_{\pi/2}^{t}\frac{\sin\mu}{\mu}\,d\mu\right|$$

$$=\left|\int_{0}^{\pi/2}\frac{\sin\mu}{\mu}\,d\mu\right|+\left|-\frac{\cos\mu}{\mu}\Big|_{\pi/2}^{t}-\int_{\pi/2}^{t}\frac{\cos\mu}{\mu^2}\,d\mu\right|$$

$$\le\left|\int_{0}^{\pi/2}\frac{\sin\mu}{\mu}\,d\mu\right|+\left|\frac{\cos t}{t}\right|+\int_{\pi/2}^{t}\frac{d\mu}{\mu^2}\le\text{const.}$$

For $0\le t\le\pi/2$ the integral $\int_{0}^{t}\frac{\sin\mu}{\mu}\,d\mu$ is also bounded in absolute value, since the integrand is a bounded function. Since the integrand is even, the latter inequality also holds for $t<0$. By Lebesgue's Bounded Convergence Theorem we have

$$\lim_{R\to\infty}\int_{-\infty}^{\infty}dP(x)\frac{1}{\pi}\int_{R(x-b)}^{R(x-a)}\frac{\sin\mu}{\mu}\,d\mu$$

$$=\int_{-\infty}^{\infty}dP(x)\lim_{R\to\infty}\frac{1}{\pi}\int_{R(x-b)}^{R(x-a)}\frac{\sin\mu}{\mu}\,d\mu.$$

We now obtain the form of this expression for different values of x.

a_1) If $x > b$ both integration bounds converge to infinity and therefore the limit of the integral is equal to zero.

a_2) If $x < a$ by an analogous argument the limit of the integral is equal to zero.

a_3) If $a < x < b$

$$\lim_{R \to \infty} \frac{1}{\pi} \int_{R(x-b)}^{R(x-a)} \frac{\sin \mu}{\mu} \, d\mu = \frac{1}{\pi} \int_{-\infty}^{\infty} \frac{\sin \mu}{\mu} \, d\mu = 1.$$

a_4) If $x = a$

$$\lim_{R \to \infty} \frac{1}{\pi} \int_{-R(b-a)}^{0} \frac{\sin \mu}{\mu} \, d\mu = \frac{1}{\pi} \int_{-\infty}^{0} \frac{\sin \mu}{\mu} \, d\mu = \frac{1}{2}.$$

a_5) If $x = b$

$$\lim_{R \to \infty} \frac{1}{\pi} \int_{0}^{R(b-a)} \frac{\sin \mu}{\mu} \, d\mu = \frac{1}{\pi} \int_{0}^{\infty} \frac{\sin \mu}{\mu} \, d\mu = \frac{1}{2}.$$

So without any assumption on the interval $[a, b]$

$$\lim_{R \to \infty} \int_{-R}^{R} \frac{e^{i\lambda a} - e^{-i\lambda b}}{i\lambda} \phi(\lambda) \, d\lambda = P\{(a, b)\} + \frac{1}{2}P\{a\} + \frac{1}{2}P\{b\}.$$

If $[a, b]$ is a continuity interval for the measure P then $P\{a\} = P\{b\} = 0$ and $P\{(a, b)\} = P\{[a, b]\}$, so the latter limit is equal to $P\{[a, b]\}$. $\quad\square$

Corollary 14.1. *If two probability measures have equal characteristic functions then they are equal.*

The proof follows from the fact that by Theorem 14.1 the distribution functions of the measures must agree at all continuity points. But if two distribution functions agree at all continuity points then they must agree at all points.

The statement in the corollary is sometimes referred to as the Unicity Theorem for characteristic functions.

One of the reasons why characteristic functions are often helpful in probability theory is that it is easy to obtain a criterion for the weak convergence of probability measures from them.

Assume that we have a sequence of probability measures $\{P_n\}$ and a probability measure P. We denote the corresponding characteristic functions by $\{\phi_n(\lambda)\}$ and $\phi(\lambda)$.

Theorem 14.2. $P_n \Rightarrow P$ *if and only if* $\lim_{n \to \infty} \phi_n(\lambda) = \phi(\lambda)$ *for any* λ.

Proof. If $P_n \Rightarrow P$, then

$$\phi_n(\lambda) = \int e^{i\lambda x}\, dP_n(x) = \int \cos \lambda x\, dP_n(x) + i \int \sin \lambda x\, dP_n(x)$$

$$\rightarrow \int \cos \lambda x\, dP(x) + i \int \sin \lambda x\, dP(x) = \int e^{i\lambda x}\, dP(x) = \phi(x),$$

so one direction in the proof of the theorem is trivial.

To prove the converse statement we establish that the family $\{P_n\}$ satisfies the conditions of Helly's Theorem. We will use the following lemma.

Lemma 14.1. *Let Q be a probability measure on the line and $\psi(\lambda)$ be its characteristic function. Then for any $\tau > 0$*

$$Q\{[-\frac{2}{\tau}, \frac{2}{\tau}]\} \geq |\frac{1}{\tau} \int_{-\tau}^{\tau} \psi(\lambda)\, d\lambda| - 1.$$

Proof. By Fubini's Theorem

$$\frac{1}{2\tau} \int_{-\tau}^{\tau} \psi(\lambda)\, d\lambda = \frac{1}{2\tau} \int_{-\tau}^{\tau} d\lambda (\int_{-\infty}^{\infty} e^{i\lambda x}\, dQ(x))$$

$$= \frac{1}{2\tau} \int_{-\infty}^{\infty} dQ(x) \int_{-\tau}^{\tau} e^{i\lambda x}\, d\lambda = \int_{-\infty}^{\infty} \frac{e^{ix\tau} - e^{-ix\tau}}{2ix\tau}\, dQ(x)$$

$$= \int_{-\infty}^{\infty} \frac{\sin x\tau}{x\tau}\, dQ(x).$$

Therefore

$$\left| \frac{1}{2\tau} \int_{-\tau}^{\tau} \psi(\lambda)\, d\lambda \right| = \left| \int_{-\infty}^{\infty} \frac{\sin x\tau}{x\tau}\, dQ(x) \right|$$

$$\leq \left| \int_{|x| \leq 2/\tau} \frac{\sin x\tau}{x\tau}\, dQ(x) \right| + \left| \int_{|x| > 2/\tau} \frac{\sin x\tau}{x\tau}\, dQ(x) \right|$$

$$\leq \int_{|x\tau| \leq 2} \left| \frac{\sin x\tau}{x\tau} \right| dQ(x) + \int_{|x\tau| > 2} \left| \frac{\sin x\tau}{x\tau} \right| dQ(x)$$

$$\leq \int_{|x\tau| \leq 2} dQ(x) + \int_{|x\tau| > 2} \left| \frac{\sin x\tau}{x\tau} \right| dQ(x) \leq Q\{[-\frac{2}{\tau}, \frac{2}{\tau}]\}$$

$$+ \frac{1}{2}(1 - Q\{[-\frac{2}{\tau}, \frac{2}{\tau}]\}) = \frac{1}{2}Q\{[-\frac{2}{\tau}, \frac{2}{\tau}]\} + \frac{1}{2}.$$

□

We now return to the proof of the theorem. Let $\phi_n(\lambda) \rightarrow \phi(\lambda)$ for each λ and let $\epsilon > 0$ be arbitrary. Since $\phi(0) = 1$ and $\phi(\lambda)$ is a continuous function there exists $\tau > 0$ such that $|\phi(\lambda) - 1| < \epsilon/4$ as soon as $|\lambda| < \tau$. Then

$$\left| \int_{-\tau}^{\tau} \phi(\lambda)\, d\lambda \right| = \left| \int_{-\tau}^{\tau} (\phi(\lambda) - 1)\, d\lambda + 2\tau \right| \geq 2\tau - \left| \int_{-\tau}^{\tau} (\phi(\lambda) - 1)\, d\lambda \right|$$

$$\geq 2\tau - \int_{-\tau}^{\tau} |\phi(\lambda) - 1|\, d\lambda > 2\tau - 2\tau\frac{\epsilon}{4} = 2\tau(1 - \frac{\epsilon}{4}).$$

Therefore

$$\left| \frac{1}{\tau} \int_{-\tau}^{\tau} \phi(\lambda) \, d\lambda \right| > 2 - \frac{\epsilon}{2}.$$

Since $\phi_n(\lambda) \to \phi(\lambda)$ and $|\phi_n(\lambda)| \leq 1$, by Lebesgue's Bounded Convergence Theorem

$$\lim_{n \to \infty} \int_{-\tau}^{\tau} \phi_n(\lambda) \, d\lambda = \int_{-\tau}^{\tau} \phi(\lambda) \, d\lambda > 2 - \frac{\epsilon}{2}.$$

So there exists an N such that for all $n \geq N$

$$\left| \frac{1}{\tau} \int_{-\tau}^{\tau} \phi_n(\lambda) \, d\lambda \right| > 2 - \epsilon.$$

By Lemma 14.1, for such n

$$P_n \left(\left[-\frac{2}{\tau}, \frac{2}{\tau} \right] \right) \geq \left| \frac{1}{\tau} \int_{-\tau}^{\tau} \phi_n(\lambda) \, d\lambda \right| - 1 > 1 - \epsilon.$$

For each $n < N$ we choose $t_n > 0$ such that $P_n([-t_n, t_n]) > 1 - \epsilon$. If we set $K = \max[\frac{2}{\tau}, \max_{1 \leq n < N} t_n]$ then we find that $P_n\{[-K, K]\} > 1 - \epsilon$ for all n. By Helly's Theorem the sequence of measures $\{P_n\}$ is weakly compact.

So we can choose a weakly convergent subsequence $\{P_{n_i}\}$, $P_{n_i} \Rightarrow \tilde{P}$. We now show that $\tilde{P} = P$. Let us denote by $\tilde{\phi}(\lambda)$ the characteristic function of $\tilde{P}$. From the first part of our theorem $\phi_n(\lambda) \to \tilde{\phi}(\lambda)$. On the other hand, by assumption $\phi_n(\lambda) \to \phi(\lambda)$. So $\phi(\lambda) = \tilde{\phi}(\lambda)$. By corollary 14.1 we have $\tilde{P} = P$.

It remains to be established that the whole sequence $P_n \Rightarrow P$. Assume that this is not true. Then for some bounded continuous function f there exists a subsequence $\{n_j\}$ such that

$$\int f(x) \, dP_{n_j}(x) \not\to \int f(x) \, dP(x).$$

We extract from the sequence $\{P_{n_j}\}$ a weakly convergent subsequence. Without loss of generality one can assume that the subsequence is the whole of $\{P_{n_j}\}$ and that $P_{n_j} \Rightarrow \bar{P}$. By the same argument as previously one can show that $\bar{P} = P$ and therefore

$$\int f(x) \, dP_{n_j}(x) \to \int f(x) \, dP(x).$$

Hence the contradiction. $\square$

Remark. One can show that if $\phi_n(\lambda) \to \phi(\lambda)$ for every λ, then this convergence is uniform on each finite interval in λ.

Lecture 15. Central Limit Theorem for Sums of Independent Random Variables

In this lecture we consider one of the most important results of probability theory, the Central Limit Theorem. We already encountered this theorem in the particular case of a sequence of Bernoulli trials in the form of the De Moivre-Laplace Limit Theorem (Lecture 3). In its simplest form the Central Limit Theorem deals with sums of independent random variables. Let us assume that a sequence of such random variables $\xi_1, \xi_2, \ldots, \xi_n$ is given. One can think of the ξ_i as the results of a sequence of independent measurements of some physical quantity. Assuming that the strong law of large numbers is applicable to the $\{\xi_i\}$ we write $\xi_i = m_i + \varsigma_i$, where $m_i = E\xi_i$, $E\varsigma_i = 0$, and we consider the arithmetic mean $\frac{1}{n}(\xi_1 + \cdots + \xi_n) = \frac{1}{n}(m_1 + \cdots + m_n) + \frac{1}{n}(\varsigma_1 + \cdots + \varsigma_n)$. It follows from the strong law of large numbers that the last term converges to zero with probability 1. The issue of the speed and type of its convergence to zero arises naturally. In the case of a sequence of measurements the answer to this question gives the precision that one can achieve with n measurements. Under some simple further assumptions it follows, from Chebyshev's inequality, that $\frac{1}{n}(\varsigma_1 + \cdots + \varsigma_n)$ takes values of order $O(1/\sqrt{n})$. Therefore we consider $\sqrt{n}(\frac{1}{n}(\varsigma_1 + \cdots + \varsigma_n)) = \frac{1}{\sqrt{n}}(\varsigma_1 + \cdots + \varsigma_n) = \eta_n$. It turns out, and this is the essence of the Central Limit Theorem, that in general, as $n \to \infty$ the random variables η_n (as functions on Ω) do not converge to a limit, but their distributions, as $n \to \infty$, have a limit which, in a well-known sense, is independent of the details of the distributions of the random variables ξ_i. We now systematically explore those considerations.

Let F_1 and F_2 be two distribution functions.

Definition 15.1. The convolution of the distribution functions F_1 and F_2 is defined to be $G(x) = \int_{-\infty}^{\infty} F_1(x - u) \, dF_2(u)$.

The convolution is written as $G = F_1 * F_2$.

Theorem 15.1. *Let ξ_1 and ξ_2 be two independent random variables with distribution functions F_1 and F_2. Then the random variable $\xi = \xi_1 + \xi_2$ has G as its distribution function.*

Proof. Let us denote by $F_{\xi_1, \xi_2}(x_1, x_2)$ the joint distribution function of the random variables ξ_1 and ξ_2. By definition

$$F_\xi(x) = P(\{\omega|\xi_1 + \xi_2 \le x\}) = \iint_{x_1+x_2 \le x} dF_{\xi_1,\xi_2}(x_1,x_2).$$

Since ξ_1 and ξ_2 are independent we have $F_{\xi_1,\xi_2}(x_1,x_2) = F_1(x_1)F_2(x_2)$ and by Fubini's Theorem

$$F_\xi(x) = \iint_{x_1+x_2 \le x} dF_1(x_1)\, dF_2(x_2) = \int dF_2(x_2)F_1(x - x_2).$$

$\square$

The formula for the convolution can be viewed as a total probability formula. If we fix a value u of the random variable ξ_2 then the probability that $\xi_1 + \xi_2 \le x$ under the condition that $\xi_2 = u$ is equal, by the independence of ξ_1 and ξ_2, to the probability that $\xi_1 \le x - u$, i.e. $F_1(x - u)$.

Corollary 15.1. $F_1 * F_2 = F_2 * F_1$.

Assume that the random variable ξ_1 takes its values in the set A_1, and that the random variable ξ_2 takes its values in the set A_2. One sometimes says that F_1 (or P_{ξ_1}) is concentrated on A_1, and F_2 (or P_{ξ_2}) is concentrated on A_2. Then G is concentrated on the set $A = A_1 + A_2$, where

$$A = \{x|x = x_1 + x_2,\ x_1 \in A_1,\ x_2 \in A_2\}.$$

The set A is called the arithmetic sum of the sets A_1 and A_2. If F_1 and F_2 are discrete then $F_1 * F_2$ is also discrete. If the distribution function F_1 has a density $p_1(x)$ then G has a density $p(x)$ and

$$p(x) = \int_{-\infty}^{\infty} p_1(x - u)\, dF_2(u).$$

Now let $\xi_1, \ldots, \xi_n$ be a sequence of n independent random variables, and $F_{\xi_1}, \ldots, F_{\xi_n}$ be their distribution functions. Then the distribution function of the sum $\varsigma_n = \xi_1 + \cdots + \xi_n$ is the n-fold convolution $F_{\varsigma_n} = F_{\xi_1} * F_{\xi_2} * \cdots * F_{\xi_n}$.

The operation of convolution, as one can see from its definition, is quite complicated. It turns out that the analysis of convolutions is significantly simplified by using characteristic functions. We denote by $\phi_{\xi_1}(\lambda)$, $\phi_{\xi_2}(\lambda), \ldots, \phi_{\xi_n}(\lambda)$ the characteristic functions of the random variables ξ_1, $\xi_2, \ldots, \xi_n$, and by $\phi_{\varsigma_n}(\lambda)$ the characteristic function of ς_n.

Theorem 15.2. $\phi_{\varsigma_n}(\lambda) = \prod_{k=1}^{n} \phi_{\xi_k}(\lambda)$.

Proof. The characteristic function $\phi_{\xi_k}(\lambda) = E \exp(i\lambda\xi_k)$. Since the random variables $\xi_1, \xi_2, \ldots, \xi_n$ are independent, the random variables $e^{i\lambda\xi_1}, \ldots, e^{i\lambda\xi_n}$, as functions of independent random variables, are also independent. This latter statement for real-valued functions can be found in Lemma 12.1. It can be extended without any modifications to complex-valued functions. Furthermore, in the case of real-valued independent random variables the expectation of their product is equal to the product of their expectations (see Lemma 12.2). This statement can also be extended, without any modifications, to the case of complex-valued random variables. Therefore

$$\phi_{\varsigma_n}(\lambda) = E e^{i\lambda\varsigma_n} = E e^{i\lambda \sum_{k=1}^{n} \xi_k} = E \prod_{k=1}^{n} e^{i\lambda\xi_k} = \prod_{k=1}^{n} \phi_{\xi_k}(\lambda).$$

□

We are now in a position to prove the Central Limit Theorem in the following important case.

Theorem 15.3. *Let $\xi_1, \xi_2, \dots \xi_n, \dots$ be a sequence of independent identically distributed random variables with distribution function F, and such that the second moment $\int_{-\infty}^{\infty} x^2 \, dF(x) = a_2 < \infty$. We denote the expectation by $m = E\xi_i = \int_{-\infty}^{\infty} x \, dF(x)$ and the variance by $d = a_2 - m^2$. Then as $n \to \infty$ the distribution of the normalized sums $\eta_n = \frac{1}{\sqrt{nd}}(\varsigma_n - nm) = \frac{1}{\sqrt{nd}}(\sum_{1}^{n} \xi_k - nm)$ converges weakly to the Gaussian normal distribution with density $\frac{1}{\sqrt{2\pi}} e^{-x^2/2}$.*

Proof. The characteristic function of the Gaussian distribution, i.e. the integral $\frac{1}{\sqrt{2\pi}} \int \exp(i\lambda x - x^2/2) \, dx = e^{-\lambda^2/2}$. The characteristic function of the random variable η_n is equal to

$$\phi_{\eta_n}(\lambda) = \left(\phi\left(\frac{\lambda}{\sqrt{nd}}\right) \exp\left(-i\frac{\lambda m}{\sqrt{nd}}\right)\right)^n, \tag{15.1}$$

where $\phi(\lambda)$ is the characteristic function of the distribution F (see property 3 of characteristic functions, Lecture 14). In order to prove the theorem it suffices to show that $\phi_{\eta_n}(\lambda) \to \exp(-\lambda^2/2)$ for every λ (Theorem 14.2).

Since $a_2 < \infty$, for any λ the integral $\int e^{i\lambda x} x^2 \, dF(x)$ exists. Therefore from the properties of the Lebesgue integral it follows that $\phi(\lambda)$ is twice differentiable, $\phi''(\lambda) = -\int e^{i\lambda x} x^2 \, dF(x)$, and $\phi''(\lambda)$ is continuous, in fact uniformly continuous. The proof of this latter property is carried out as the proof of the uniform continuity of characteristic functions (see Property 5 of characteristic functions, Lecture 14). Therefore for small λ

$$\phi(\lambda) = 1 + im\lambda - \frac{\lambda^2}{2} a_2 + o(\lambda^2). \tag{15.2}$$

The last term calls for a further remark. We write the real and imaginary parts of the characteristic function $\phi(\lambda)$ separately and for each part we write a Taylor expansion with a Lagrange remainder up to the second order. The intermediate point is in general different for each part. By the continuity of $\phi''(\lambda)$ and since $\phi''(0) = -a_2$, the second derivative of the real part converges to $-a_2$ and that of the imaginary part converges to zero. This leads to equation (15.2). If we substitute it into (15.1) we obtain

$$\phi_{\eta_n}(\lambda) = \left[\left(1 + \frac{i\lambda m}{\sqrt{nd}} - \frac{\lambda^2 a_2}{2nd} + o\left(\frac{\lambda^2}{nd}\right)\right)\left(1 - \frac{i\lambda m}{\sqrt{nd}} - \frac{\lambda^2 m^2}{2nd} + o\left(\frac{\lambda^2}{nd}\right)\right)\right]^n$$

$$= \left(1 + \frac{i\lambda m}{\sqrt{nd}} - \frac{i\lambda m}{\sqrt{nd}} - \frac{\lambda^2 a_2}{2nd} - \frac{\lambda^2 m^2}{2nd} + \frac{\lambda^2 m^2}{nd} + o\left(\frac{\lambda^2}{nd}\right)\right)^n$$

$$= \left(1 - \frac{\lambda^2}{2nd}(a_2 - m^2) + o\left(\frac{1}{n}\right)\right)^n$$

$$= \left(1 - \frac{\lambda^2}{2n} + o\left(\frac{1}{n}\right)\right)^n \to e^{-\lambda^2/2}$$

for any fixed λ. $\qquad\qquad\qquad\qquad\qquad\qquad\qquad\qquad\qquad\qquad\qquad$ $\square$

We now quote, without proof, a more general formulation of the Central Limit Theorem which deals with sequences of independent random variables with different distributions.

Theorem 15.4 (Lyapunov). *Let $\{\xi_k\}$ be a sequence of independent random variables, $m_k = E\xi_k$, $d_k = \mathrm{Var}\,\xi_k$ and $E|\xi_k - m_k|^3 = c_k^3$. If $(1/D_n^{3/2})\sum_{k=1}^n c_k$ $\to 0$ as $n \to \infty$, where $D_n = \sum_{k=1}^n d_k$, then the probability distribution of the normed sums $\eta_n = (1/\sqrt{D_n})(\sum_{k=1}^n \xi_k - \sum_{k=1}^n m_k)$ converges weakly to the Gaussian distribution density $(1/\sqrt{2\pi})e^{-x^2/2}$.*

We recall that $D_n = \mathrm{Var}(\sum_{k=1}^n \xi_k)$, since for independent random variables the variance of the sum is equal to the sum of the variances. A more general form of the Central Limit Theorem is that with Lindeberg's condition, which is almost necessary. Let F_k be the distribution function of the random variable ξ_k. We assume that Lindeberg's condition holds, namely for each $\epsilon > 0$

$$\frac{1}{D_n}\sum_{k=1}^n \int_{x:|x-m_k|\geq\epsilon\sqrt{D_n}} (x - m_k)^2\,dF_k(x) \to 0,\ n \to \infty.$$

Theorem 15.5. *If Lindeberg's condition is satisfied then the limiting distribution (as $n \to \infty$) of the normed sum η_n is Gaussian. If in addition a condition of asymptotic smallness is satisfied:*

$$\max_{1\leq k\leq n} P\{|\frac{\xi_k - m_k}{\sqrt{D_n}}| \geq \epsilon\} \to 0,$$

then Lindeberg's condition is necessary for the convergence of distribution of η_n to the Gaussian distribution.

There are quite a few generalizations of the Central Limit Theorem, where the condition of independence of the random variables is replaced by a condition of weak dependence in one sense or another. Other important generalizations are those connected with vector-valued random variables.

The appearance and the universality of the Gaussian distribution in the Central Limit Theorem can seem rather enigmatic and surprising. The goal of the following remarks is to elucidate both those circumstances. It will be convenient to consider a sequence of numbers $n_p = 2^p$ and assume that $E\xi_i = 0$. We define the random variable $\gamma_p = 2^{-p/2} \sum_{k=1}^{2^p} \xi_k$, where the ξ_k are identically distributed and have a finite second moment. Then $\gamma_{p+1} = (1/\sqrt{2})(\gamma_p' + \gamma_p'')$ where $\gamma_p' = 2^{-p/2} \sum_{k=1}^{2^p} \xi_k$ and $\gamma_p'' = 2^{-p/2} \sum_{k=2^p+1}^{2^{p+1}} \xi_k$. We see that γ_p' and γ_p'' are independent, identically distributed random variables. If F_p is the distribution function of γ_p then

$$F_{p+1}(x) = \int_{-\infty}^{\infty} F_p(x\sqrt{2} - u) \, dF_p(u). \tag{15.3}$$

So the sequence F_p can be obtained from F_0 by iterations of a non-linear operation obtained from transformation (15.3), which we denote by T. This means that if for any distribution function F we set $TF = \int_{-\infty}^{\infty} F(x\sqrt{2} - u) \, dF(u)$ we have $F_{p+1} = TF_p$ and $F_p = T^p F_0$. It is easy to check directly that the Gaussian distribution function $G(x) = (1/\sqrt{2\pi}) \int_{-\infty}^{x} e^{-u^2/2} \, du$ is a fixed point of the transformation T, i.e. $G = TG$. The fact that for a wide class of distribution functions F_0 the convergence $F_p \Rightarrow G$ holds is related to the stability of this fixed point. We now note that if we set $d(F) = \int x^2 \, dF(x)$ for a distribution function F such that $\int x \, dF(x) = 0$, i.e. $d(F)$ is the variance of the distribution F, then

$$d(TF) = d(F). \tag{15.4}$$

This follows from the fact that the variance of the random variable $\gamma = (1/\sqrt{2})(\gamma' + \gamma'')$, where γ' and γ'' are independent, identically distributed random variables, is the same as the variance of γ' and γ''. Equation (15.4) shows that the variance is the "first integral" of the non-linear transformation T.

In the general theory of non-linear operators the investigation of the stability of a fixed point begins with an investigation of its stability with respect to the linear approximation. In our case the problem on linear stability arises in the following way. Set $F_\epsilon = \int_{-\infty}^{x} (1 + \epsilon h(u)) \frac{1}{\sqrt{2\pi}} \exp(-\frac{u^2}{2}) \, du$ and write

$$TF_\epsilon = \int_{-\infty}^{x} (1 + \epsilon L h) \frac{1}{\sqrt{2\pi}} e^{-\frac{1}{2}u^2} \, du + o(\epsilon).$$

The linear operator L which appears in the last expression is the linearization of the operator T at the point G. It is easy to write it explicitly as

$$Lh = \frac{\sqrt{2}}{\sqrt{\pi}} \int_{-\infty}^{\infty} h\left(u + \frac{x}{\sqrt{2}}\right) e^{-\frac{1}{2}u^2} \, du.$$

The integral operator that we have written is sometimes called the Gaussian integral operator. One can show that it is self-adjoint for a suitable choice of a scalar product. We now find its spectrum and eigenfunctions. One can show directly that the eigenfunctions of L are the Hermite polynomials $h =$

$P_k(x)$ (see Lecture 9), which correspond to the eigenvalues $\mu_k = 2^{1-k/2}$, $k = 0, 1, 2, 3, \ldots$. We see that $\mu_0, \mu_1 > 1$, $\mu_2 = 1$, and all the remaining eigenvalues are less than 1. In other words, in directions which correspond to P_0 and P_1 the operator T is non-stable, while in the direction which corresponds to P_2 it is marginal and in all remaining directions it is stable. We note here that we must consider perturbations h for which

$$\int h(u)e^{-\frac{u^2}{2}}\,du = 0, \quad \int uh(u)e^{-\frac{u^2}{2}}\,du = 0.$$

The first condition corresponds to the fact that the perturbation of G is such that $\int_{-\infty}^{x}(1+\epsilon h(u))e^{-u^2/2}\,du$ remains a distribution function, and the second condition means that the expectation of the perturbed distribution function remains equal to zero. Thus the perturbations are orthogonal to the non-stable directions. Furthermore since the variance $d(F)$ is the first integral of the transformation T then the perturbations must leave the variance unchanged:

$$\int_{-\infty}^{\infty} u^2 h(u)e^{-\frac{u^2}{2}}\,du = 0,$$

i.e. h must be orthogonal to the second Hermite polynomial as well. In the rest of the subspace L has a spectrum which is strictly less than 1. This shows that the fixed point G is stable for the linear approximation with respect to perturbations which lie in the class of distribution functions with zero expectation and fixed variances.

We can now conclude from the general theory of non-linear operators that for distribution functions F with zero expectation and $d(F) = 1$ which lie in a sufficiently small neighborhood of G, we have $T^p(F) \Rightarrow G$. The formulations of the Central Limit Theorem given above mean that G is a stable point not only for a small, but also for a large, class of distributions.

A suitable generalization of this argument makes it possible to deal with the problem of constructing limit distributions for sequences of dependent random variables. However we will not deal with this matter in any more detail.

For a large class of problems it is necessary to consider slowly decreasing distributions for which $\int x^2\,dF(x) = \infty$. It turns out that this is where new limit distributions emerge. We now consider a particular case.

Theorem 15.6. *Let* $\xi_1, \xi_2, \ldots, \xi_n, \ldots$ *be a sequence of independent, identically distributed random variables for which the distribution function F has a density $p(x)$ such that*

i) $p(x) = p(-x)$; *and*
ii) $p(x) \sim c/(|x|^{\alpha+1})$ *for some* $\alpha \in (0,2)$, *where c is a constant.*

Then the distribution of the normed sum $\eta_n = n^{-1/\alpha}(\xi_1 + \cdots + \xi_n)$ *as* $n \to \infty$ *converges to a limit distribution whose characteristic function is of the form* $\exp(-c_1|\lambda|^\alpha)$ *for some constant* $c_1 > 0$.

The theory of limit distributions for sums of independent random variables is one of the most developed branches of probability theory, in which many remarkable results are known.

Lecture 16. Probabilities of Large Deviations

At the beginning of this course on probability theory we considered the probabilities $P(|\sum_{k=1}^{n} \xi_k - \sum_{k=1}^{n} m_k| \geq t)$ with $m_k = E\xi_k$ for a sequence of independent random variables $\xi_1, \xi_2, \ldots$, and we estimated those probabilities by Chebyshev's inequality:

$$P\left(\left|\sum_{k=1}^{n} \xi_k - \sum_{k=1}^{n} m_k\right| \geq t\right) \leq \frac{1}{t^2} \sum_{k=1}^{n} d_k, \quad d_k = \text{Var}\xi_k.$$

It follows in particular that if the random variables ξ_i are identically distributed then for some constant c which does not depend on n, and with $d = d_k$,

a_1) for $t = c\sqrt{n}$ we have d/c^2 on the right hand side;
a_2) for $t = cn$ we have $d/c^2 n$ on the right hand side.

We know from the Central Limit Theorem that in case a_1) as $n \to \infty$ the corresponding probability converges to a positive limit which can be calculated by using the Gaussian distribution. This means that in a_1) the order of magnitude of the estimates obtained from Chebyshev's inequality is correct. On the other hand in the case a_2) the estimate from Chebyshev's inequality is often too large, and very crude. In this lecture we investigate the possibility of obtaining more precise estimates for the probability in case a_2).

Let us consider a sequence of independent identically distributed random variables. We denote their common distribution functions by F. We make the following basic assumption about F: the integral

$$R(\lambda) = \int_{-\infty}^{\infty} e^{\lambda x} \, dF(x) < \infty \tag{16.1}$$

for all λ, $-\infty < \lambda < \infty$. This condition is automatically satisfied if all the ξ_i are bounded: $|\xi_i| \leq c = \text{const}$. It is also satisfied if the probabilities of large values of the ξ_i decrease faster than exponentially.

We now investigate several properties of the function $R(\lambda)$. From the finiteness of the integral in (16.1) it follows that the derivatives

$$R'(\lambda) = \int_{-\infty}^{\infty} x e^{\lambda x} \, dF(x), \quad R''(\lambda) = \int_{-\infty}^{\infty} x^2 e^{\lambda x} \, dF(x)$$

exist for all λ. Let us consider $m(\lambda) = R'(\lambda)/R(\lambda)$. Then

$$m'(\lambda) = \frac{R''(\lambda)}{R(\lambda)} - \left(\frac{R'(\lambda)}{R(\lambda)}\right)^2 = \int_{-\infty}^{\infty} \frac{x^2}{R(\lambda)} e^{\lambda x}\, dF(x) - \left(\int_{-\infty}^{\infty} \frac{x}{R(\lambda)} e^{\lambda x}\, dF(x)\right)^2.$$

We construct a new distribution function $F_\lambda(x) = \frac{1}{R(\lambda)} \int_{(-\infty,x]} e^{\lambda x}\, dF(x)$ for each λ. Then $m(\lambda) = \int x\, dF_\lambda(x)$ is the expectation, computed with respect to this distribution, and $m'(\lambda)$ is the variance. Therefore $m'(\lambda) > 0$ if F is a non-trivial distribution, i.e. is not concentrated at a point. We exclude this latter case from our further considerations. Since $m'(\lambda) > 0$, $m(\lambda)$ is a monotonically increasing function. We will need some information on $\lim m(\lambda)$ as $\lambda \to \pm\infty$.

We say that M^+ is an upper limit in probability for the random variable ξ, if $P(\xi > M^+) = 0$, and $P(M^+ - \epsilon \le \xi \le M^+) > 0$ for any $\epsilon > 0$. One can define a concept of lower limit in probability M^- in an analogous way. If $P(\xi > M) > 0$ $(P(\xi < M) > 0)$ for any M, then $M^+ = \infty$ $(M^- = -\infty)$. In all remaining cases M^+ and M^- are finite.

Lemma 16.1. $\lim_{\lambda \to \infty} m(\lambda) = M^+$, $\lim_{\lambda \to -\infty} m(\lambda) = M^-$.

Proof. We prove only the first statement, since the second is proved in an analogous way. Assume first that $M^+ = \infty$. We show that $m(\lambda) \ge M$ for any preassigned M if $\lambda > \lambda(M)$. We have

$$R'(\lambda) = \int_{-\infty}^{\infty} x e^{\lambda x}\, dF(x) \ge \int_{2M}^{\infty} x e^{\lambda x}\, dF(x).$$

Furthermore

$$R(\lambda) = \int_{-\infty}^{\infty} e^{\lambda x}\, dF(x) = \int_{-\infty}^{2M} e^{\lambda x}\, dF(x) + \int_{2M}^{\infty} e^{\lambda x}\, dF(x)$$

$$\le e^{2\lambda M} + \int_{2M}^{\infty} e^{\lambda x}\, dF(x) = \int_{2M}^{\infty} e^{\lambda x}\, dF(x)\left(1 + \frac{e^{2\lambda M}}{\int_{2M}^{\infty} e^{\lambda x}\, dF(x)}\right).$$

Since $M^+ = \infty$, for any M there exists a half-open interval with $(a, a+1]$, $a > 4M$ such that $F(a+1) - F(a) > 0$. Therefore

$$\int_{2M}^{\infty} e^{\lambda x}\, dF(x) \ge \int_{(a,a+1]} e^{\lambda x}\, dF(x) \ge e^{a\lambda}\left(F(a+1) - F(a)\right)$$

$$\ge e^{4M\lambda}\left(F(a+1) - F(a)\right),$$

and

$$\frac{e^{2\lambda M}}{\int_{2M}^{\infty} e^{\lambda x}\, dF(x)} \le \frac{e^{2\lambda M}}{e^{4\lambda M}\left(F(a+1) - F(a)\right)} = \frac{e^{-2\lambda M}}{F(a+1) - F(a)}.$$

For $\lambda > \lambda(M)$ the latter expression is less than 1. For such λ

$$\frac{R'(\lambda)}{R(\lambda)} \ge \frac{\int_{2M}^{\infty} x e^{\lambda x}\, dF(x)}{\left(\int_{2M}^{\infty} e^{\lambda x}\, dF(x)\right)\left(1 + \frac{e^{\lambda M}}{\int_{2M}^{\infty} e^{\lambda x}\, dF(x)}\right)}$$

$$\ge \frac{1}{2} \int_{2M}^{\infty} x \frac{e^{\lambda x}\, dF(x)}{\int_{2M}^{\infty} e^{\lambda x}\, dF(x)} \ge \frac{2M}{2} = M.$$

To obtain the last inequality we used the fact that $\frac{e^{\lambda x}\, dF(x)}{\int_{2M}^{\infty} e^{\lambda x}\, dF(x)}$ generates a probability measure on the half-line $(2M, \infty)$. So the lemma is proved for the case where $M^+ = \infty$.

We now turn to the case of a finite M^+. We show that for any $\delta > 0$ there exists a $\lambda(\delta)$ such that $m(\lambda) \geq M^+ - \delta$ if $\lambda > \lambda(\delta)$. As previously

$$R'(\lambda) = \int_{-\infty}^{\infty} x e^{\lambda x}\, dF(x) \geq \int_{M^+ - \delta/2}^{\infty} x e^{\lambda x}\, dF(x),$$

$$R(\lambda) = \int_{-\infty}^{M^+ - \delta/2} e^{\lambda x}\, dF(x) + \int_{M^+ - \delta/2}^{\infty} e^{\lambda x}\, dF(x)$$

$$\leq e^{\lambda(M^+ - \frac{1}{2}\delta)} + \int_{M^+ - \delta/2}^{\infty} e^{\lambda x}\, dF(x)$$

$$= \int_{M^+ - \delta/2}^{\infty} e^{\lambda x}\, dF(x)\Big(1 + \frac{e^{\lambda(M^+ - \delta/2)}}{\int_{M^+ - \delta/2}^{\infty} e^{\lambda x}\, dF(x)}\Big).$$

Since M^+ is an upper limit in probability we have $F(M^+) - F(M^+ - \frac{1}{2}\delta) > 0$ for any $\delta > 0$. Therefore

$$\frac{e^{\lambda(M^+ - \delta/2)}}{\int_{M^+ - \delta/2}^{\infty} e^{\lambda x}\, dF(x)} \leq \frac{e^{\lambda(M^+ - \delta/2)}}{\int_{(M^+ - \delta/4, M^+]} e^{\lambda x}\, dF(x)}$$

$$\leq \frac{e^{\lambda(M^+ - \delta/2)}}{e^{\lambda(M^+ - \delta/4)}\big(F(M^+) - F(M^+ - \frac{1}{4}\delta)\big)}$$

$$= e^{-\frac{\lambda}{4}\delta}\, \frac{1}{F(M^+) - F(M^+ - \frac{1}{4}\delta)}.$$

Therefore

$$\frac{R'(\lambda)}{R(\lambda)} \geq \frac{\int_{M^+ - \delta/2}^{\infty} x e^{\lambda x}\, dF(x)}{\int_{M^+ - \delta/2}^{\infty} e^{\lambda x}\, dF(x)}\, \frac{1}{\Big(1 + \frac{e^{-\lambda\delta/4}}{F(M^+) - F(M^+ - \frac{1}{4}\delta)}\Big)}.$$

The first quotient, by the same arguments as above, is greater than or equal to $M^+ - \frac{1}{2}\delta$. The second quotient can be made arbitrarily close to 1 by an appropriate choice of λ. □

We can now turn directly to an estimate of the probabilities of interest to us. Let c be fixed, and set $m = E\xi_i < c < M^+$. We consider the probability $P_{n,c} = P(\xi_1 + \cdots + \xi_n > cn)$. Since $c > m$ the probability we are investigating is that of the random variables taking values far away from their mathematical expectation. Such values are called large deviations (from the expectation). Let λ_0 be such that $m(\lambda_0) = c$. Note that $m = m(0) < c$. Therefore $\lambda_0 > 0$ by the monotonicity of $m(\lambda_0)$.

Theorem 16.1. $P_{n,c} \leq B_n \left(R(\lambda_0) e^{-\lambda_0 c} \right)^n$ where $\lim_{n \to \infty} B_n = \frac{1}{2}$.

Proof. We have

$$P_{n,c} = \int \cdots \int_{x_1 + \cdots + x_n > cn} dF(x_1) \ldots dF(x_n)$$

$$\leq (R(\lambda_0))^n \, e^{-\lambda_0 cn} \int \cdots \int_{x_1 + \cdots + x_n > cn} \frac{e^{-\lambda_0(x_1 + \cdots + x_n)}}{(R(\lambda_0))^n} dF(x_1) \ldots dF(x_n)$$

$$= (R(\lambda_0) e^{-\lambda_0 c})^n \int \cdots \int_{x_1 + \cdots + x_n > cn} dF_{\lambda_0}(x_1) \ldots dF_{\lambda_0}(x_n).$$

The latter integral is the probability, computed with respect to the distribution F_{λ_0}, that $\xi_1 + \cdots + \xi_n > cn$. But the distribution F_{λ_0} was chosen in such a way that $E_{\lambda_0} \xi_i = c$. Therefore

$$\int \cdots \int_{x_1 + \cdots + x_n > cn} dF_{\lambda_0}(x_1) \ldots dF_{\lambda_0}(x_n) = P_{\lambda_0} \{ \xi_1 + \cdots + \xi_n > cn \}$$

$$= P_{\lambda_0} \{ \xi_1 + \cdots + \xi_n - nm(\lambda_0) > 0 \}$$

$$= P_{\lambda_0} \{ \frac{\xi_1 + \cdots + \xi_n - nm(\lambda_0)}{\sqrt{nd(\lambda_0)}} > 0 \} \to \frac{1}{2}$$

as $n \to \infty$. Here we have denoted by $d(\lambda_0)$ the variance of the random variables computed with respect to the distribution F_{λ_0}. $\qquad\square$

The lower estimate turns out to be somewhat worse.

Theorem 16.2. *For any $b > 0$ there exists $p(b) = p > 0$ such that*

$$P_{n,c} \geq (R(\lambda_0) e^{-\lambda_0 c})^n \, e^{-\lambda_0 b \sqrt{n}} \, p_n,$$

with

$$\lim_{n \to \infty} p_n = p(b) > 0.$$

Proof. As in Theorem 16.1

$$P_{n,c} \geq \int \cdots \int_{cn < x_1 + \cdots + x_n < cn + b\sqrt{n}} dF(x_1) \ldots dF(x_n)$$

$$\geq (R(\lambda_0))^n \, e^{-\lambda_0(cn + b\sqrt{n})}$$

$$\int \cdots \int_{cn < x_1 + \cdots + x_n < cn + b\sqrt{n}} e^{-\lambda(x_1 + \cdots + x_n)} \frac{dF(x_1)}{R(\lambda_0)} \cdots \frac{dF(x_n)}{R(\lambda_0)}$$

$$= (R(\lambda_0) e^{-\lambda_0 c})^n \, e^{-\lambda_0 b \sqrt{n}} \int \cdots \int_{cn < x_1 + \cdots + x_n < cn + b\sqrt{n}} dF_{\lambda_0}(x_1) \cdots dF_{\lambda_0}(x_n).$$

The latter integral, as in the case of Theorem 16.1, converges to a positive limit by the Central Limit Theorem.   □

In Theorems 16.1 and 16.2 the number $R(\lambda_0)e^{-\lambda_0 c} = r(\lambda_0)$ is involved. It is clear that $r(0) = 1$. We show that $r(\lambda_0) < 1$ for all $\lambda_0 \neq 0$. We have $\ln r(\lambda_0) = \ln R(\lambda_0) - \lambda_0 c = -(\ln R(\lambda_0) - \ln R(0)) - \lambda_0 c$. By Taylor's formula we have:

$$\ln R(0) - \ln R(\lambda_0) = -(\ln R(\lambda_0))'\lambda_0 + \frac{\lambda_0^2}{2}(\ln R(\lambda_1))'',$$

where λ_1 is an intermediate point between 0 and λ_0. Furthermore

$$(\ln R(\lambda_0))' = \frac{R'(\lambda_0)}{R(\lambda_0)} = m(\lambda_0) = c, \quad \text{and} \quad (\ln R(\lambda_1))'' > 0$$

since it is the variance of the random variable ξ_i computed with respect to the distribution F_{λ_1}. Thus

$$\ln r(\lambda_0) = -\frac{\lambda_0^2}{2}(\ln R(\lambda_1))'' < 0.$$

From Theorems 16.1 and 16.2 we obtain:

Corollary 16.1. $\lim_{n \to \infty} \frac{1}{n} \ln P_{n,c} = \ln r(\lambda_0) < 0.$

Indeed let $b = 1$ in Theorem 16.2. Then

$$\ln r(\lambda_0) - \frac{\lambda_0}{\sqrt{n}} - \frac{\ln p_n}{n} \leq \frac{\ln P_{n,c}}{n} \leq \ln r(\lambda_0) + \frac{1}{n}\ln B_n.$$

Passing to the limit as $n \to \infty$ completes the proof.   □

Corollary 16.1 shows that the probabilities $P_{n,c}$ decrease exponentially in n, i.e. much faster than was given by Chebyshev's inequality.

Printing: Weihert-Druck GmbH, Darmstadt
Binding: Theo Gansert Buchbinderei GmbH, Weinheim

I. Karatzas, Columbia University, New York, NY;
S. E. Shreve, Carnegie Mellon University,
Pittsburgh, PA

Brownian Motion and Stochastic Calculus

2nd ed. 1991. XXIII, 470 pp. 10 figs.
(Graduate Texts in Mathematics, Vol. 113)
Softcover ISBN 3-540-97655-8

Designed as a text for graduate courses in stochastic processes and written for readers familiar with measure-theoretic probability and discrete-time processes who wish to explore stochastic processes in continuous time. The vehicle chosen for this exposition is Brownian motion, which is presented as the canonical example of both a martingale and a Markov process with continuous paths. In this context, the theory of stochastic integration and stochastic calculus is developed. The power of this calculus is illustrated by results concerning representations of martingales and change of measure on Wiener space, and this in turn permits a presentation of recent advances in financial economics (options pricing and consumption/ investment optimization).

The book contains a detailed discussion of weak and strong solutions of stochastic differential equations and a study of local time for semimartingales, with special emphasis on the theory of Brownian local time. The text is complemented by a large number of problems and exercises.

Springer-Verlag
Berlin
Heidelberg
New York
London
Paris
Tokyo
Hong Kong
Barcelona
Budapest

D. Revuz, University of Paris VII;
M. Yor, University Pierre et Marie Curie

Continuous Martingales and Brownian Motion

1991. IX, 533 pp. 8 figs. (Grundlehren der
mathematischen Wissenschaften, Bd. 293)
Hardcover ISBN 3-540-52167-4

This work provides a detailed study of Brownian
Motion, via the Itô stochastic calculus of continuous
processes, e.g. diffusions, continuous semimartingales:
it should facilitate the reading and understanding of
research papers in this area, and be of interest both to
graduate students and to more advanced readers, either
working primarily with stochastic processes, or doing
research in an area involving stochastic processes,
e.g. mathematical physics, and economics.
The emphasis is on methods, rather than generality.
After a first introductory chapter, each of the subsequent ones introduces a new method or idea,
e.g. stochastic integration, local times, excursions,
weak convergence, and describes its applications
to Brownian motion; some of
these appear for the first time in
book form. One of the important features of the book is the
large number of exercises
which, at the same time, give
additional results and will help
the reader master the subject
more easily.

Springer-Verlag
Berlin
Heidelberg
New York
London
Paris
Tokyo
Hong Kong
Barcelona
Budapest